Chapter 7
Division Facts and Strategies

Houghton
Mifflin
Harcourt

Made in the United States
Text printed on 100%
recycled paper

Houghton Mifflin Harcourt

Printed in the U.S.A.

ISBN 978-0-544-34213-2

2 3 4 5 6 7 8 9 10 0928 22 21 20 19 18 17 16 15 14

4500476012 ^ B C D E F G

Dear Students and Families,

Welcome to **Go Math!**, Grade 3! In this exciting mathematics program, there are hands-on activities to do and real-world problems to solve. Best of all, you will write your ideas and answers right in your book. In **Go Math!**, writing and drawing on the pages helps you think deeply about what you are learning, and you will really understand math!

By the way, all of the pages in your **Go Math!** book are made using recycled paper. We wanted you to know that you can Go Green with **Go Math!**

Sincerely,

The Authors

Made in the United States
Text printed on 100% recycled paper

GO MATH!

Authors

Juli K. Dixon, Ph.D.
Professor, Mathematics Education
University of Central Florida
Orlando, Florida

Edward B. Burger, Ph.D.
President, Southwestern University
Georgetown, Texas

Steven J. Leinwand
Principal Research Analyst
American Institutes for Research (AIR)
Washington, D.C.

Matthew R. Larson, Ph.D.
K-12 Curriculum Specialist for Mathematics
Lincoln Public Schools
Lincoln, Nebraska

Martha E. Sandoval-Martinez
Math Instructor
El Camino College
Torrance, California

Contributor

Rena Petrello
Professor, Mathematics
Moorpark College
Moorpark, California

English Language Learners Consultant

Elizabeth Jiménez
CEO, GEMAS Consulting
Professional Expert on English Learner Education
Bilingual Education and Dual Language
Pomona, California

Place Value and Operations with Whole Numbers

Critical Area

Critical Area Developing understanding of multiplication and division and strategies for multiplication and division within 100

7 Division Facts and Strategies 363

COMMON CORE STATE STANDARDS

3.OA Operations and Algebraic Thinking

Cluster A Represent and solve problems involving multiplication and division.
3.OA.A.3, 3.OA.A.4

Cluster C Multiply and divide within 100.
3.OA.C.7

Cluster D Solve problems involving the four operations, and identify and explain patterns in arithmetic.
3.OA.D.8

GO DIGITAL

Go online! Your math lessons are interactive. Use *i*Tools, Animated Math Models, the Multimedia eGlossary, and more.

Chapter 7 Overview

In this chapter, you will explore and discover answers to the following **Essential Questions**:

- What strategies can you use to divide?
- How can you use a related multiplication fact to divide?
- How can you use factors to divide?
- What types of problems can be solved by using division?

CRITICAL AREA REVIEW PROJECT INVENTING TOYS: www.thinkcentral.com

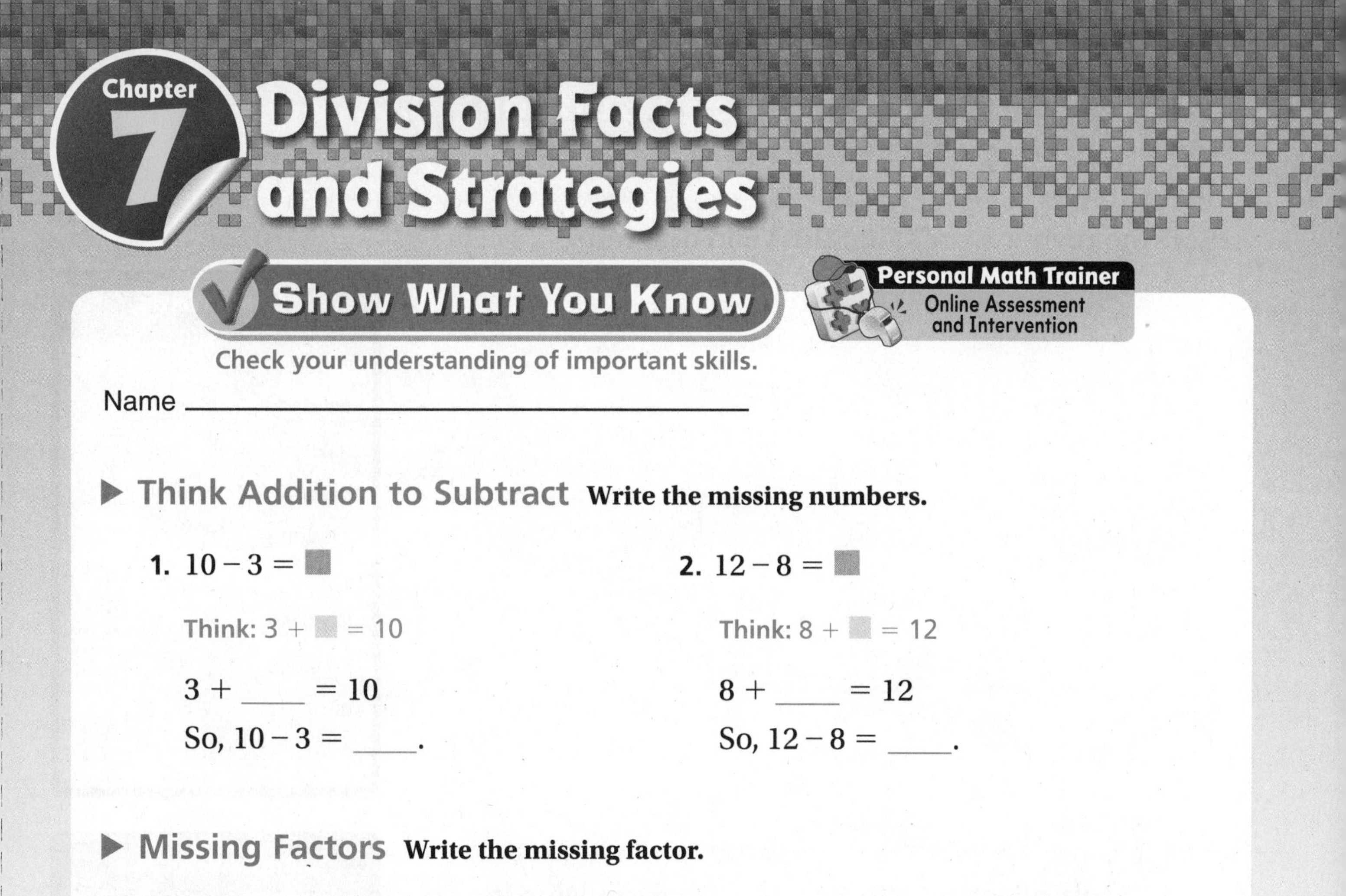

Chapter 7 Division Facts and Strategies

Show What You Know

Personal Math Trainer
Online Assessment and Intervention

Check your understanding of important skills.

Name ______________________

▶ Think Addition to Subtract Write the missing numbers.

1. $10 - 3 = \blacksquare$

Think: $3 + \blacksquare = 10$

$3 +$ ____ $= 10$

So, $10 - 3 =$ ____.

2. $12 - 8 = \blacksquare$

Think: $8 + \blacksquare = 12$

$8 +$ ____ $= 12$

So, $12 - 8 =$ ____.

▶ Missing Factors Write the missing factor.

3. $2 \times$ ____ $= 10$

4. $42 =$ ____ $\times 7$

5. ____ $\times 6 = 18$

▶ Multiplication Facts Through 9 Find the product.

6. ____ $= 6 \times 9$

7. $3 \times 8 =$ ____

8. $4 \times 4 =$ ____

On Monday, the students in Mr. Carson's class worked in pairs. On Tuesday, the students worked in groups of 3. On Wednesday, the students worked in groups of 4. Each day the students made equal groups with no student left out of a group. Help students to find how many students could be in Mr. Carson's class.

Vocabulary Builder

▶ Visualize It

Sort the review words into the Venn diagram.

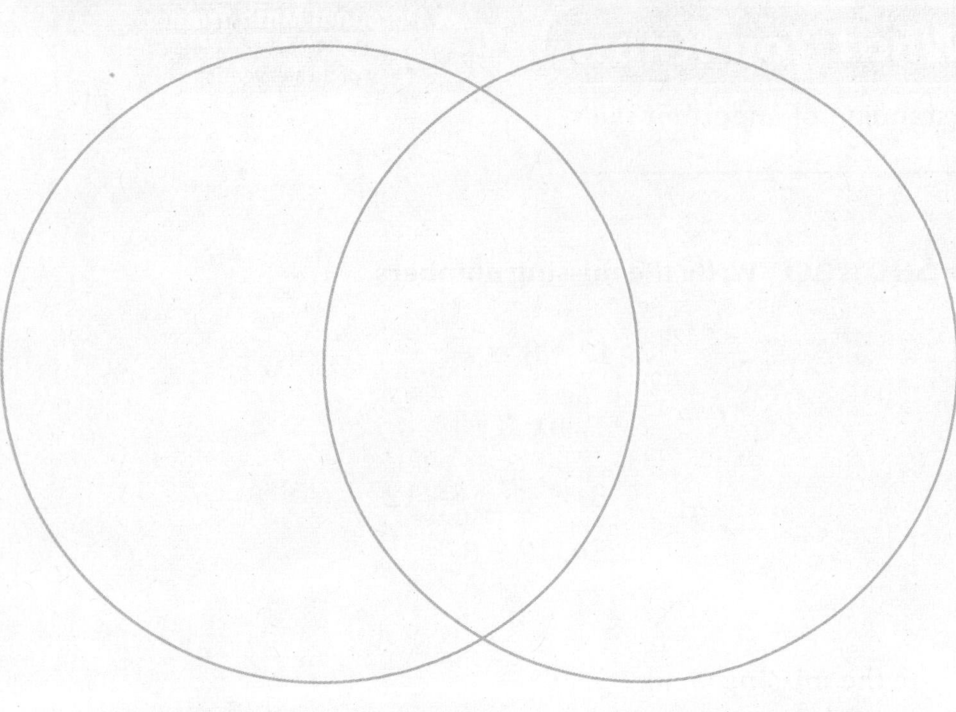

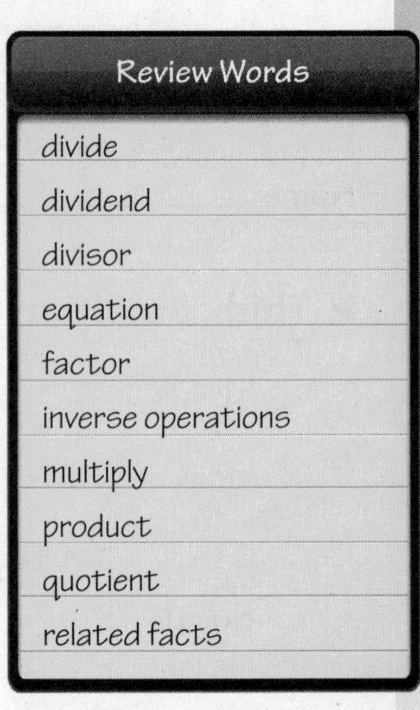

Preview Words

order of operations

▶ Understand Vocabulary

Complete the sentences by using the review and preview words.

1. An ______________ is a number sentence that uses the equal sign to show that two amounts are equal.

2. The ________________ is a special set of rules that gives the order in which calculations are done to solve a problem.

3. ______________ are a set of related multiplication and division equations.

• Interactive Student Edition
• Multimedia *e*Glossary

Chapter 7 Vocabulary

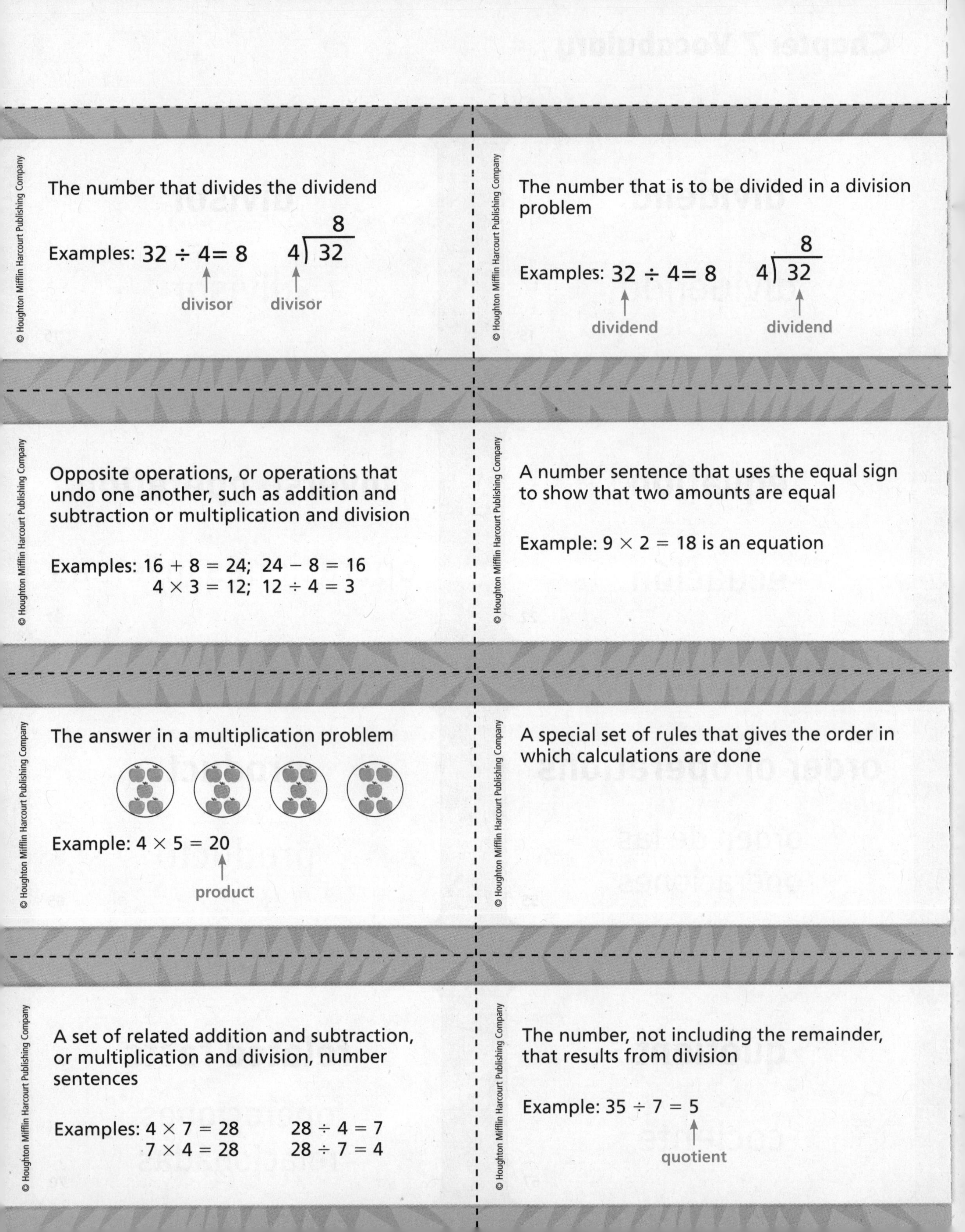

The number that divides the dividend

Examples: $32 \div 4 = 8$ $\quad 4\overline{)32}$ with quotient 8

The number that is to be divided in a division problem

Examples: $32 \div 4 = 8$ $\quad 4\overline{)32}$ with quotient 8

Opposite operations, or operations that undo one another, such as addition and subtraction or multiplication and division

Examples: $16 + 8 = 24$; $24 - 8 = 16$
$4 \times 3 = 12$; $12 \div 4 = 3$

A number sentence that uses the equal sign to show that two amounts are equal

Example: $9 \times 2 = 18$ is an equation

The answer in a multiplication problem

Example: $4 \times 5 = 20$

A special set of rules that gives the order in which calculations are done

A set of related addition and subtraction, or multiplication and division, number sentences

Examples: $4 \times 7 = 28$ $\quad 28 \div 4 = 7$
$7 \times 4 = 28$ $\quad 28 \div 7 = 4$

The number, not including the remainder, that results from division

Example: $35 \div 7 = 5$

Concentration

Word Box
- dividend
- divisor
- equation
- inverse operations
- order of operations
- product
- quotient
- related facts

For 2–3 players

Materials

- 1 set of word cards

How to Play

1. Put the cards face-down in rows. Take turns to play.
2. Choose two cards and turn them face-up.
   - If the cards show a word and its meaning. it's a match. Keep the pair and take another turn.
   - If the cards do not match, turn them back over.
3. The game is over when all cards have been matched. The players count their pairs. The player with the most pairs wins.

The Write Way

Reflect

Choose one idea. Write about it.

- Do $16 \div 8$ and $8 \div 16$ have the same quotient? Explain why or why not.
- Explain the Order of Operations in your own words.
- Write a creative story that includes division by 2, 5, or 10.

Name ____________________

Divide by 2

Essential Question What does dividing by 2 mean?

Common Core **Operations and Algebraic Thinking—3.OA.A.3** *Also 3.OA.A.2, 3.OA.C.7*
MATHEMATICAL PRACTICES
MP4, MP5, MP6

Unlock the Problem Real World

There are 10 hummingbirds and 2 feeders in Marissa's backyard. If there are an equal number of birds at each feeder, how many birds are at each one?

- What do you need to find?

____

- Circle the numbers you need to use.
- What can you use to help solve the problem? ____________________

Activity 1

Use counters to find how many in each group.

Materials ■ counters ■ MathBoard

MODEL

- Use 10 counters.
- Draw 2 circles on your MathBoard.
- Place 1 counter at a time in each circle until all 10 counters are used.
- Draw the rest of the counters to show your work.

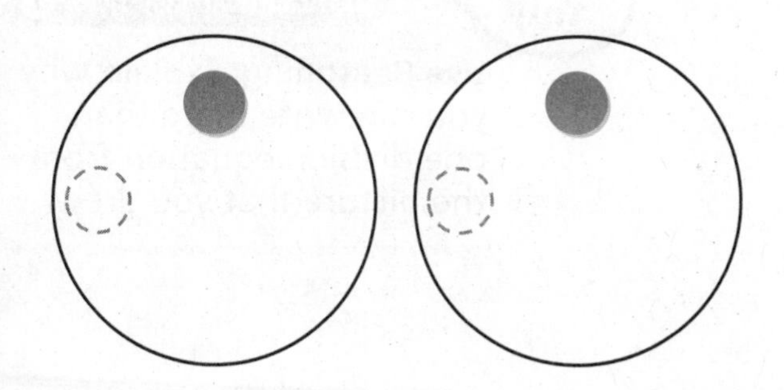

THINK

_____ in all

_____ equal groups

_____ in each group

RECORD

$10 \div 2 = 5$ or $2\overline{)10}$ with 5 above

Read: Ten divided by two equals five.

There are _____ counters in each of the 2 groups.

So, there are _____ hummingbirds at each feeder.

A hummingbird can fly right, left, up, down, forward, backward, and even upside down! ▶

Math Talk MATHEMATICAL PRACTICES 1

Analyze What does each number in $10 \div 2 = 5$ represents from the word problem?

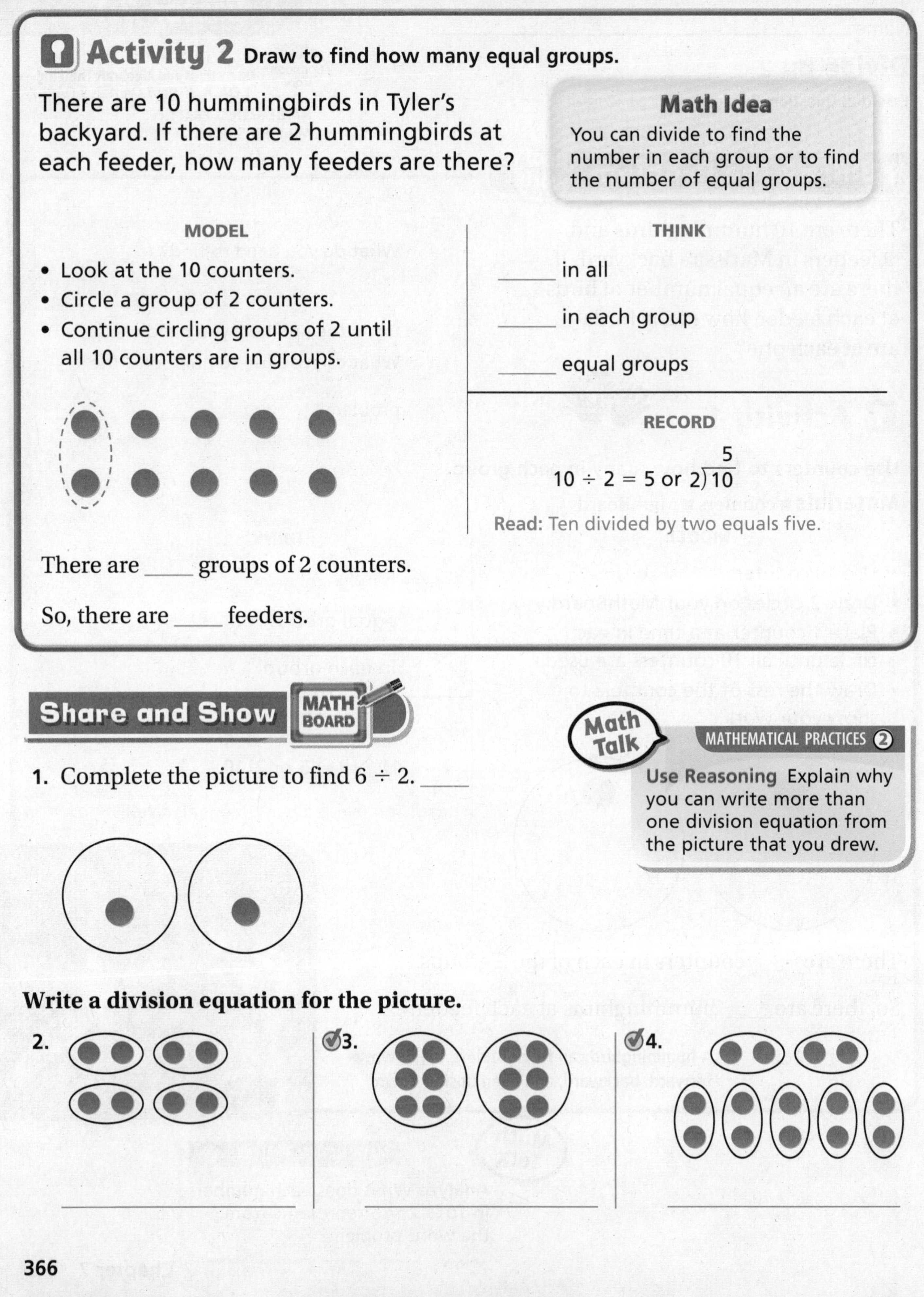

Activity 2 Draw to find how many equal groups.

There are 10 hummingbirds in Tyler's backyard. If there are 2 hummingbirds at each feeder, how many feeders are there?

Math Idea

You can divide to find the number in each group or to find the number of equal groups.

MODEL

- Look at the 10 counters.
- Circle a group of 2 counters.
- Continue circling groups of 2 until all 10 counters are in groups.

There are ____ groups of 2 counters.

So, there are ____ feeders.

THINK

____ in all

____ in each group

____ equal groups

RECORD

$10 \div 2 = 5$ or $2\overline{)10}$ with quotient 5

Read: Ten divided by two equals five.

Share and Show MATH BOARD

1. Complete the picture to find $6 \div 2$. ____

Math Talk MATHEMATICAL PRACTICES 2

Use Reasoning Explain why you can write more than one division equation from the picture that you drew.

Write a division equation for the picture.

2. ____

3. ____

4. ____

Name ____________________

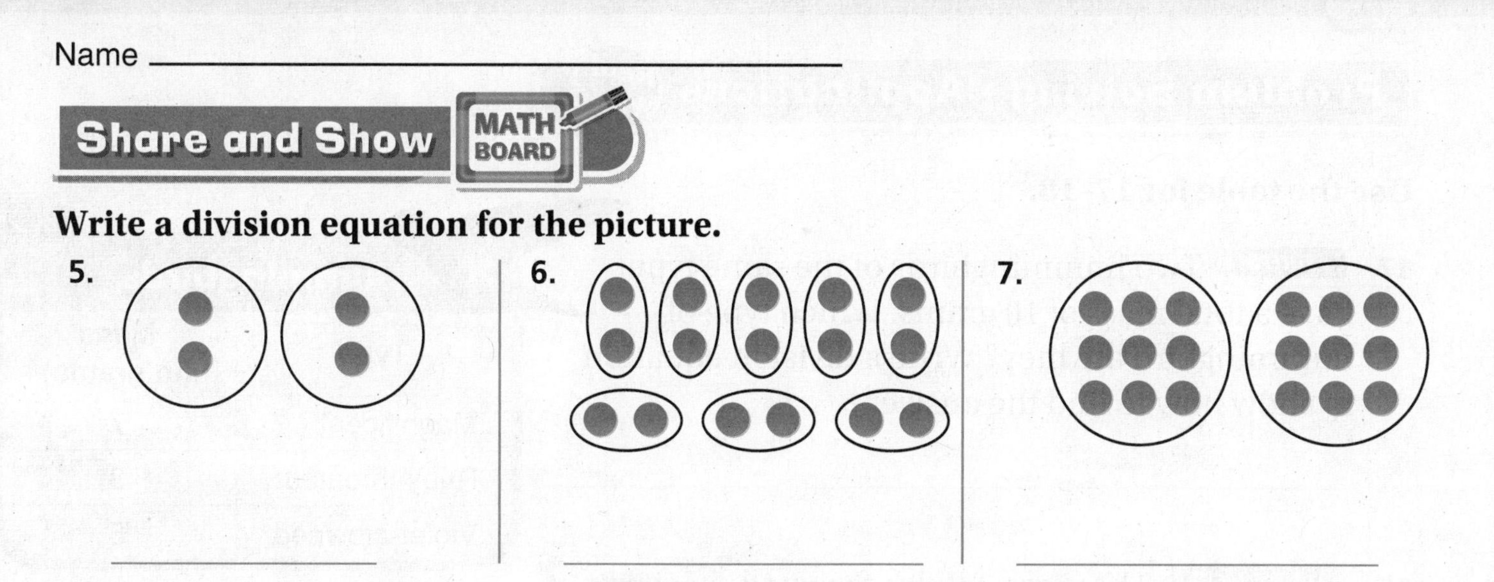

Write a division equation for the picture.

5. ____________

6. ____________

7. ____________

Find the quotient. You may want to draw a quick picture to help.

8. $2 \div 2 =$ ____

9. $16 \div 2 =$ ____

10. $2\overline{)20}$

MATHEMATICAL PRACTICE 2 **Reason Abstractly** **Algebra** **Find the unknown number.**

11. ____ $\div 2 = 5$

12. ____ $\div 2 = 2$

13. ____ $\div 2 = 3$

14. ____ $\div 2 = 8$

15. Lin makes a tile design with 24 tiles. Half are red and half are blue. He takes away 4 red tiles. How many red tiles are in the design now?

____

16. GO DEEPER Becky made 2 video tapes while a giant hummingbird fed 4 times and a ruby throated hummingbird fed 8 times from her new feeder. Each video tape caught the same number of feedings by the hummingbirds. How many feedings were shown on each video tape? Justify your answer.

____

____

Problem Solving • Applications Real World

Use the table for 17–18.

Hummingbirds	
Type	Mass (in grams)
Magnificent	7
Ruby-throated	3
Violet-crowned	5

17. GO DEEPER Two hummingbirds of the same type have a total mass of 10 grams. Which type of hummingbird are they? Write a division equation to show how to find the answer.

____

18. THINK SMARTER There are 3 Ruby-throated hummingbirds and 2 of another type of hummingbird at a feeder. The birds have a mass of 23 grams in all. What other type of hummingbird is at the feeder? **Explain.**

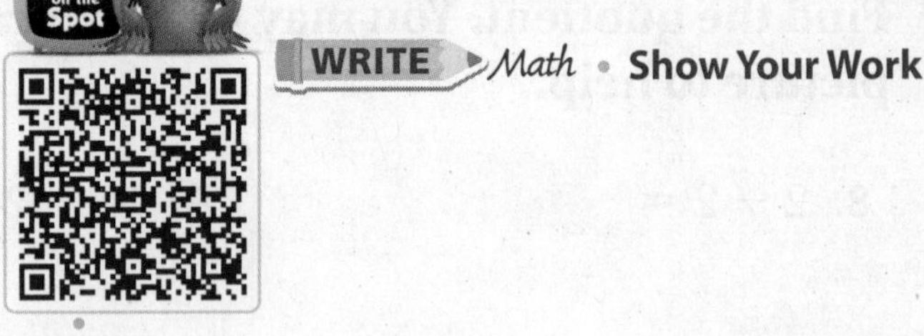

____

____

____

____

____

19. THINK SMARTER Ryan has 18 socks.

Divisor	Quotient
○ 1	○ 1
○ 2	○ 3
○ 6	○ 9
○ 18	○ 18

Select one number from each column to show the division equation represented by the picture.

$18 \div \frac{?}{\text{(divisor)}} = \frac{?}{\text{(quotient)}}$

Name ______________________________

Divide by 2

COMMON CORE STANDARD—3.OA.A.3
Represent and solve problems involving multiplication and division.

Write a division equation for the picture.

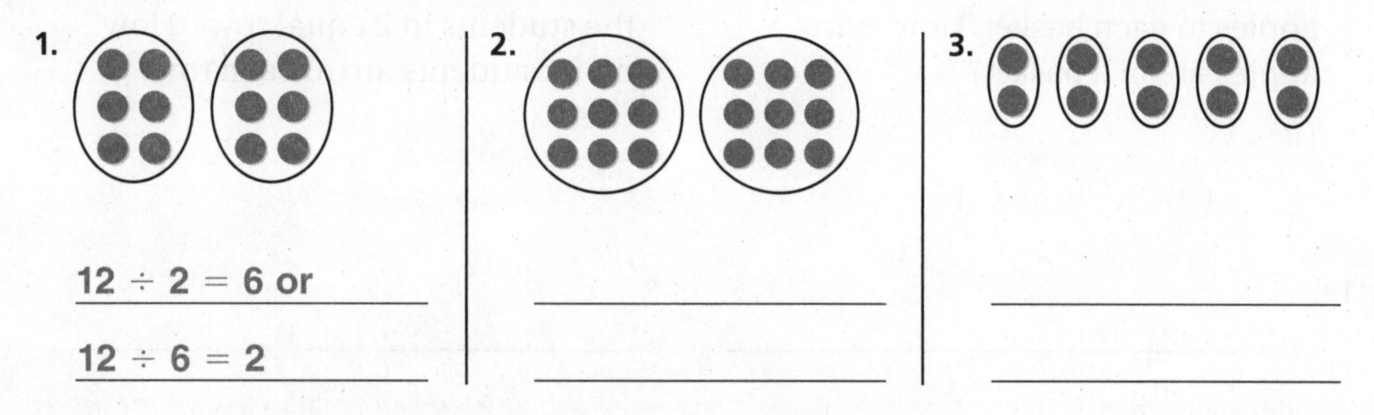

1. $12 \div 2 = 6$ or $12 \div 6 = 2$

2. ____________ ____________

3. ____________ ____________

Find the quotient. You may want to draw a quick picture to help.

4. _____ $= 14 \div 2$

5. $2\overline{)18}$

6. $16 \div 2 =$ _____

Problem Solving Real World

7. Mr. Reynolds, the gym teacher, divided a class of 16 students into 2 equal teams. How many students were on each team?

____

8. Sandra has 10 books. She divides them into groups of 2 each. How many groups can she make?

____

9. WRITE Math Explain how to divide an amount by 2. Use the terms *dividend, divisor,* and *quotient.*

____

____

Lesson Check (3.OA.A.3)

1. Ava has 12 apples and 2 baskets. She puts an equal number of apples in each basket. How many apples are in a basket?

2. There are 8 students singing a song in the school musical. Ms. Lang put the students in 2 equal rows. How many students are in each row?

Spiral Review (3.OA.A.2, 3.OA.A.3, 3.OA.D.9)

3. Find the product.

$$2 \times 6$$

4. Jayden plants 24 trees. He plants the trees equally in 3 rows. How many trees are in each row?

5. Describe the pattern below.

9, 12, 15, 18, 21, 24

6. A tricycle has 3 wheels. How many wheels are there on 4 tricycles?

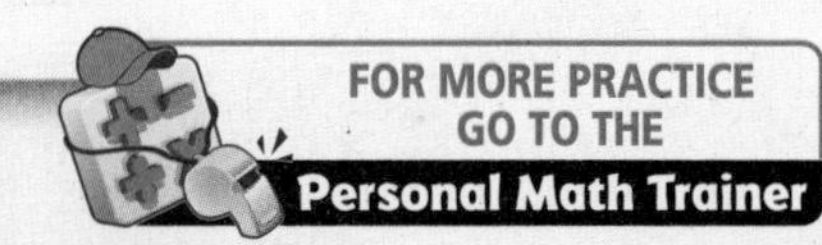

Name ______________________________

Divide by 10

Essential Question What strategies can you use to divide by 10?

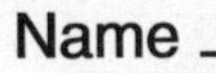

Common Core **Operations and Algebraic Thinking—3.OA.C.7** *Also 3.OA.A.2, 3.OA.A.3, 3.OA.A.4, 3.OA.B.6*

MATHEMATICAL PRACTICES
MP1, MP2, MP5, MP8

Unlock the Problem (Real World)

There are 50 students going on a field trip to the Philadelphia Zoo. The students are separated into equal groups of 10 students each. How many groups of students are there?

- What do you need to find?

____

- Circle the numbers you need to use.

One Way Use repeated subtraction.

- Start with 50.
- Subtract 10 until you reach 0.
- Count the number of times you subtract 10.

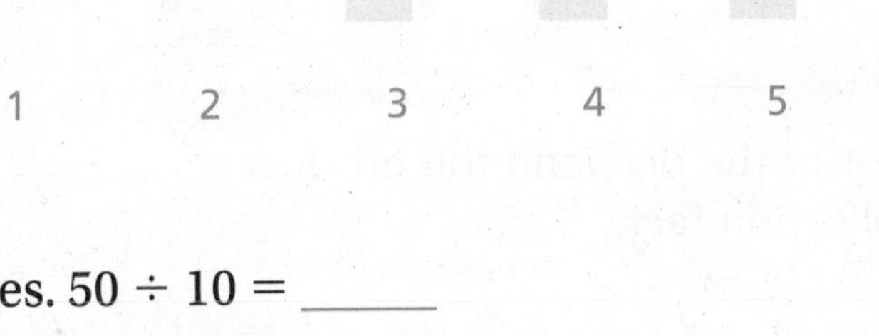

You subtracted 10 five times. $50 \div 10 =$ ______

So, there are ______ groups of 10 students.

Other Ways

A **Use a number line.**

- Start at 50 and count back by 10s until you reach 0.
- Count the number of times you jumped back 10.

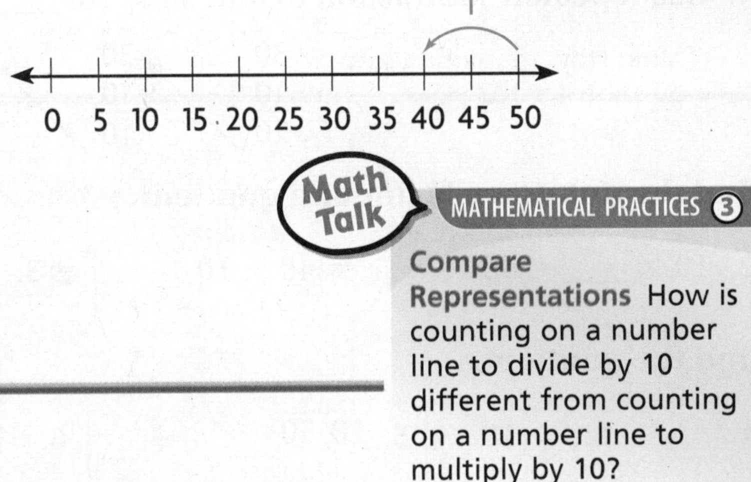

You jumped back by 10 five times.

$50 \div 10 =$ ______

Math Talk MATHEMATICAL PRACTICES 3

Compare Representations How is counting on a number line to divide by 10 different from counting on a number line to multiply by 10?

B **Use a multiplication table.**

Divide. 50 ÷ 10 = ■

Since division is the opposite of multiplication, you can use a multiplication table to find a quotient.

Think of a related multiplication fact.

■ × 10 = 50

STEP 1 Find the factor, 10, in the top row.

STEP 2 Look down to find the product, 50.

STEP 3 Look left to find the unknown factor, ____.

×	0	1	2	3	4	5	6	7	8	9	10
0	0	0	0	0	0	0	0	0	0	0	0
1	0	1	2	3	4	5	6	7	8	9	10
2	0	2	4	6	8	10	12	14	16	18	20
3	0	3	6	9	12	15	18	21	24	27	30
4	0	4	8	12	16	20	24	28	32	36	40
5	0	5	10	15	20	25	30	35	40	45	50
6	0	6	12	18	24	30	36	42	48	54	60
7	0	7	14	21	28	35	42	49	56	63	70
8	0	8	16	24	32	40	48	56	64	72	80
9	0	9	18	27	36	45	54	63	72	81	90
10	0	10	20	30	40	50	60	70	80	90	100

Since ____ × 10 = 50, then 50 ÷ 10 = ____.

In Step 1, is the divisor or the dividend the given factor in the related multiplication fact?

____

In Step 2, is the divisor or the dividend the product in the related multiplication fact?

____

The quotient is the unknown factor.

Share and Show MATH BOARD

Math Talk MATHEMATICAL PRACTICES 3

Compare Strategies What are some other strategies besides repeated subtraction to solve 30 ÷ 10?

1. Use repeated subtraction to find 30 ÷ 10. ____

Think: How many times do you subtract 10?

$$\begin{array}{r}30\\-10\\\hline 20\end{array} \nearrow \begin{array}{r}20\\-10\\\hline 10\end{array} \nearrow \begin{array}{r}10\\-10\\\hline ■\end{array}$$

Find the unknown factor and quotient.

2. 10 × ____ = 40 ____ = 40 ÷ 10

3. 10 × ____ = 60 60 ÷ 10 = ____

Find the quotient.

4. ____ = 20 ÷ 10

5. $10\overline{)50}$

6. $10\overline{)70}$

7. 90 ÷ 10 = ____

Name ______________________________

On Your Own

Find the unknown factor and quotient.

8. $10 \times \_\_\_ = 70 \quad 70 \div 10 = \_\_\_$

9. $10 \times \_\_\_ = 10 \quad 10 \div 10 = \_\_\_$

Find the quotient.

10. $50 \div 10 = \_\_\_$

11. $\_\_\_ = 60 \div 10$

12. $10\overline{)40}$

13. $10\overline{)80}$

MATHEMATICAL PRACTICE 2 **Reason Quantitatively** **Algebra** **Write <, >, or =.**

14. $10 \div 1 \bigcirc 4 \times 10$

15. $17 - 6 \bigcirc 18 \div 2$

16. $4 \times 4 \bigcirc 8 + 8$

17. $23 + 14 \bigcirc 5 \times 8$

18. $70 \div 10 \bigcirc 23 - 16$

19. $9 \times 0 \bigcirc 9 + 0$

20. GO DEEPER There are 70 pieces of chalk in a box. If each of 10 students gets an equal number of chalk pieces, how many pieces of chalk does each student get?

____

21. GO DEEPER Elijah wrote his name in 15 school shirts. Cora wrote her name in 15 school shirts. Together they labeled 10 shirts each day. On how many days did Elijah and Cora label shirts?

____

22. GO DEEPER Peyton has 32 cubes. Myra has 18 cubes. If both students use all of the cubes to make trains with 10 cubes, how many trains can they make?

____

Problem Solving • Applications Real World

Use the picture graph for 23–25.

Animal Stickers	
Elephants	▢ ▢ ▢ ▯
Giraffes	▢ ▢ ▯
Monkeys	▢ ▢ ▢ ▢
Key: Each ▢ = 10 stickers.	

23. Lyle wants to add penguins to the picture graph. There are 30 stickers of penguins. How many symbols should Lyle draw for penguins?

____

24. GO DEEPER Write a word problem using information from the picture graph. Then solve your problem.

____

____

WRITE Math • Show Your Work

25. THINK SMARTER **Sense or Nonsense?** Lena wants to put the monkey stickers in an album. She says she will use more pages if she puts 5 stickers on a page instead of 10 stickers on a page. Is she correct? Explain.

____

____

____

26. MATHEMATICAL PRACTICE 6 **Explain** how a division problem is like an unknown factor problem.

____

____

____

27. THINK SMARTER Lilly found 40 seashells. She put 10 seashells in each bucket. How many buckets did Lilly use? Show your work.

_____ buckets

Name ______________________

Divide by 10

Practice and Homework
Lesson 7.2

Common Core **COMMON CORE STANDARD—3.OA.C.7**
Multiply and divide within 100.

Find the unknown factor and quotient.

1. $10 \times$ 2 $= 20$ $20 \div 10 =$ 2

2. $10 \times$ ____ $= 70$ $70 \div 10 =$ ____

3. $10 \times$ ____ $= 80$ $80 \div 10 =$ ____

4. $10 \times$ ____ $= 30$ $30 \div 10 =$ ____

Find the quotient.

5. $60 \div 10 =$ ____

6. ____ $= 40 \div 4$

7. $20 \div 2 =$ ____

8. $50 \div 10 =$ ____

9. $10\overline{)40}$

10. $10\overline{)70}$

11. $10\overline{)100}$

12. $10\overline{)20}$

Problem Solving Real World

13. Pencils cost 10¢ each. How many pencils can Brent buy with 90¢?

____

14. Mrs. Marks wants to buy 80 pens. If the pens come in packs of 10, how many packs does she need to buy?

____

15. WRITE Math Write and solve a word problem that involves dividing by 10.

____

____

Lesson Check (3.OA.C.7)

1. Gracie uses 10 beads on each necklace she makes. She has 60 beads to use. How many necklaces can Gracie make?

2. A florist arranges 10 flowers in each vase. How many vases does the florist need to arrange 40 flowers?

Spiral Review (3.OA.A.2, 3.OA.A.3, 3.OA.A.4, 3.NBT.A.3)

3. What is the unknown factor?

$$7 \times p = 14$$

4. Aspen Bakery sold 40 boxes of rolls in one day. Each box holds 6 rolls. How many rolls did the bakery sell?

5. Mr. Samuels buys a sheet of stamps. There are 4 rows with 7 stamps in each row. How many stamps does Mr. Samuels buy?

6. There are 56 students going on a field trip to the science center. The students tour the center in groups of 8. How many groups of students are there?

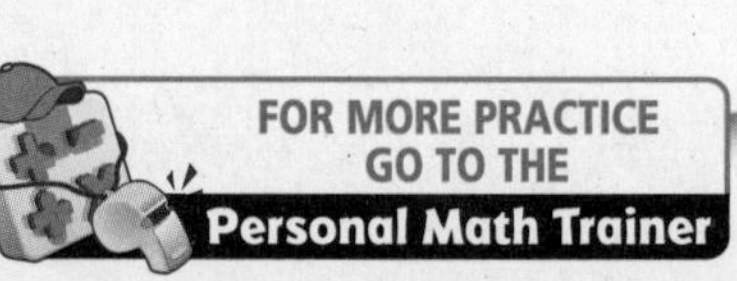

Name ______________________________

Divide by 5

Essential Question What does dividing by 5 mean?

Common Core **Operations and Algebraic Thinking—3.OA.A.3** *Also 3.OA.A.2, 3.OA.C.7*

MATHEMATICAL PRACTICES
MP1, MP2, MP5, MP7

Unlock the Problem (Real World)

Kaley wants to buy a new cage for Coconut, her guinea pig. She has saved 35¢. If she saved a nickel each day, for how many days has she been saving?

- How much is a nickel worth?

  ____

One Way Count up by 5s.

- Begin at 0.
- Count up by 5s until you reach 35. 5, 10, ____, ____, ____, ____, ____
- Count the number of times you count up. 1 2 3 4 5 6 7

You counted up by 5 seven times. $35 \div 5 =$ ____

So, Kaley has been saving for ____ days.

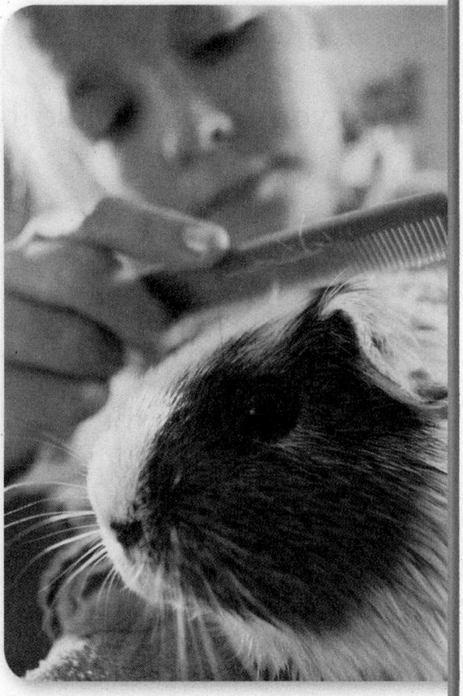

Another Way (Hands On)

Count back on a number line.

- Start at 35.
- Count back by 5s until you reach 0. Complete the jumps on the number line.
- Count the number of times you jumped back 5.

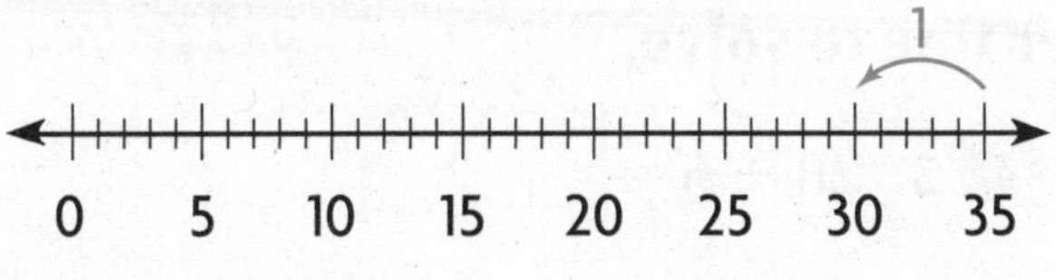

Number line: 0 5 10 15 20 25 30 35 (jump 1 from 35 to 30)

You jumped back by 5 ________ times.

$35 \div 5 =$ ____

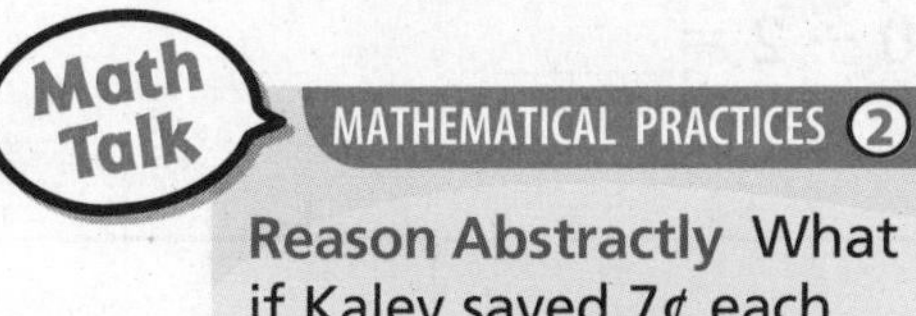

Math Talk MATHEMATICAL PRACTICES 2

Reason Abstractly What if Kaley saved 7¢ each day instead of a nickel? What would you do differently to find how many days she has saved?

Strategies for Multiplying and Dividing with 5

You have learned how to use doubles to multiply. Now you will learn how to use doubles to divide by 5.

Use 10s facts, and then take half to multiply with 5.

When one factor is 5, you can use a 10s fact. $5 \times 2 = \blacksquare$

First, multiply by 10. $10 \times 2 =$ ____

After you multiply, take half of the product. $20 \div 2 =$ ____

So, $5 \times 2 =$ ____.

Divide by 10, and then double to divide by 5.

When the divisor is 5 and the dividend is even, you can use a 10s fact. $30 \div 5 = \blacksquare$

First, divide by 10. $30 \div 10 =$ ____

After you divide, double the quotient. $3 +$ ____ $=$ ____

So, $30 \div 5 =$ ____.

Share and Show

MATH BOARD

1. Count back on the number line to find $15 \div 5$. ____

Math Talk

MATHEMATICAL PRACTICES 6

Explain how counting up to solve a division problem is like counting back on a number line.

Use count up or count back on a number line to solve.

2. $10 \div 2 =$ ____

3. $20 \div 5 =$ ____

Find the quotient.

4. $50 \div 5 =$ ____

5. $5 \div 5 =$ ____

6. $45 \div 5 =$ ____

Name ______________________________

On Your Own

Use count up or count back on a number line to solve.

7. $30 \div 5 =$ ____

8. $25 \div 5 =$ ____

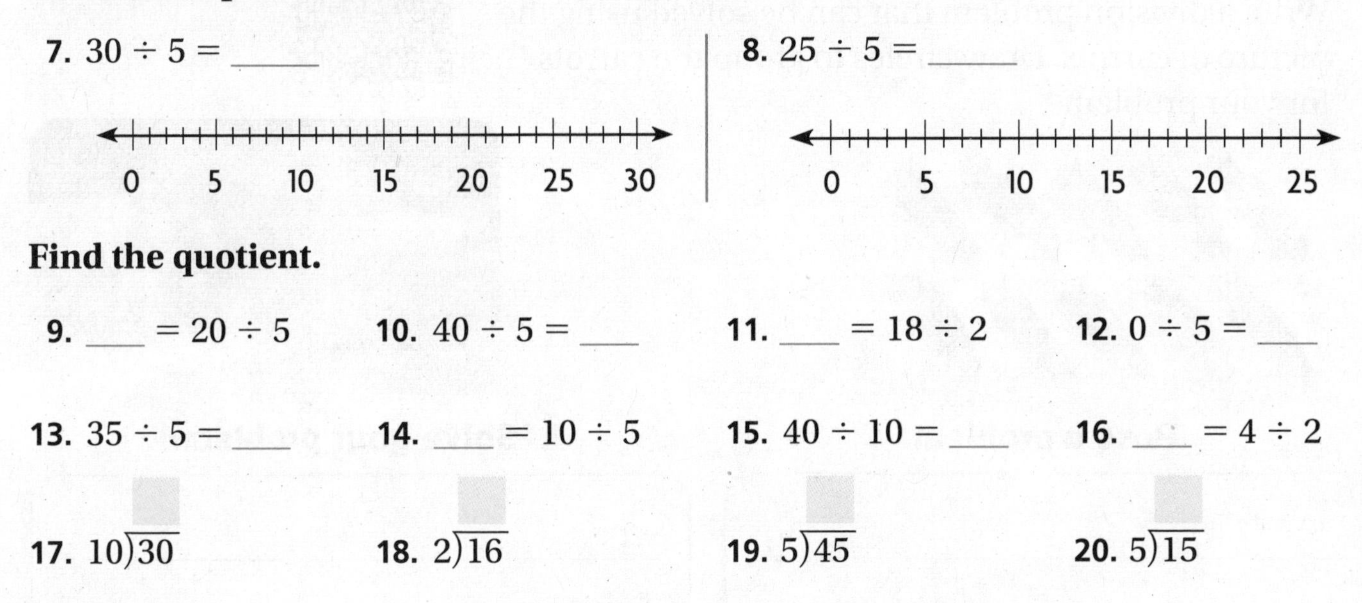

Find the quotient.

9. ____ $= 20 \div 5$

10. $40 \div 5 =$ ____

11. ____ $= 18 \div 2$

12. $0 \div 5 =$ ____

13. $35 \div 5 =$ ____

14. ____ $= 10 \div 5$

15. $40 \div 10 =$ ____

16. ____ $= 4 \div 2$

17. $10\overline{)30}$

18. $2\overline{)16}$

19. $5\overline{)45}$

20. $5\overline{)15}$

MATHEMATICAL PRACTICE 7 **Look for a Pattern** **Algebra** **Complete the table.**

21.

×	1	2	3	4	5
10					
5					

22.

÷	10	20	30	40	50
10					
5					

Problem Solving • Applications Real World

23. MATHEMATICAL PRACTICE 1 **Evaluate** Guinea pigs eat hay, pellets, and vegetables. If Wonder Hay comes in a 5-pound bag and costs $15, how much does 1 pound of hay cost?

____

24. GO DEEPER Ana picks 25 apples. Pedro picks 20 apples. Ana and Pedro use the apples to make apple pies. They put 5 apples in each pie. How many apple pies do they make?

____

25. GO DEEPER The clerk at the pet supply store works 45 hours a week. He works an equal number of hours on Monday through Friday. He works an extra 5 hours on Saturday. How many hours does he work on each weekday?

____

26. THINK SMARTER **Pose a Problem** Maddie went to a veterinary clinic. She saw the vet preparing some carrots for the guinea pigs.

Write a division problem that can be solved using the picture of carrots. Draw circles to group the carrots for your problem.

Pose a problem. **Solve your problem.**

- Group the carrots in a different way. Then write a problem for the new groups. Solve your problem.

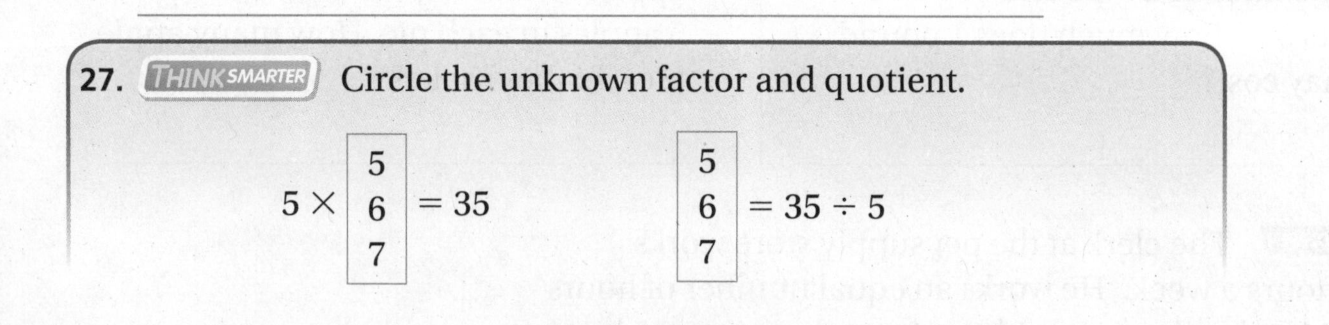

27. THINK SMARTER Circle the unknown factor and quotient.

$5 \times$ [5 / 6 / 7] $= 35$

[5 / 6 / 7] $= 35 \div 5$

Name ______________________

Divide by 5

Common Core **COMMON CORE STANDARD—3.OA.A.3**
Represent and solve problems involving multiplication and division.

Use count up or count back on a number line to solve.

1. $40 \div 5 =$ <u>8</u>

0 5 10 15 20 25 30 35 40

2. $25 \div 5 =$ ____

0 5 10 15 20 25

Find the quotient.

3. ____ $= 10 \div 5$

4. ____ $= 30 \div 5$

5. $14 \div 2 =$ ____

6. $5 \div 5 =$ ____

7. ____ $= 0 \div 5$

8. $20 \div 5 =$ ____

9. $25 \div 5 =$ ____

10. ____ $= 35 \div 5$

11. $5\overline{)20}$

12. $10\overline{)70}$

13. $5\overline{)15}$

14. $5\overline{)40}$

Problem Solving Real World

15. A model car maker puts 5 wheels in each kit. A machine makes 30 wheels at a time. How many packages of 5 wheels can be made from the 30 wheels?

____

16. A doll maker puts a small bag with 5 hair ribbons inside each box with a doll. How many bags of 5 hair ribbons can be made from 45 hair ribbons?

____

17. WRITE Math Write about which method you prefer to use to divide by 5—counting up, counting back on a number line, or dividing by 10, and then doubling the quotient. Explain why.

____

____

Lesson Check (3.OA.A.3)

1. A model train company puts 5 boxcars with each train set. How many sets can be completed using 35 boxcars?

2. A machine makes 5 buttons at a time. Each doll shirt gets 5 buttons. How many doll shirts can be finished with 5 buttons?

Spiral Review (3.OA.A.3, 3.MD.B.4)

3. Julia earns $5 each day running errands for a neighbor. How much will Julia earn if she runs errands for 6 days in one month?

4. Marcus has 12 slices of bread. He uses 2 slices of bread for each sandwich. How many sandwiches can Marcus make?

Use the line plot for 5–6.

5. How many students have no pets?

6. How many students answered the question "How many pets do you have?"

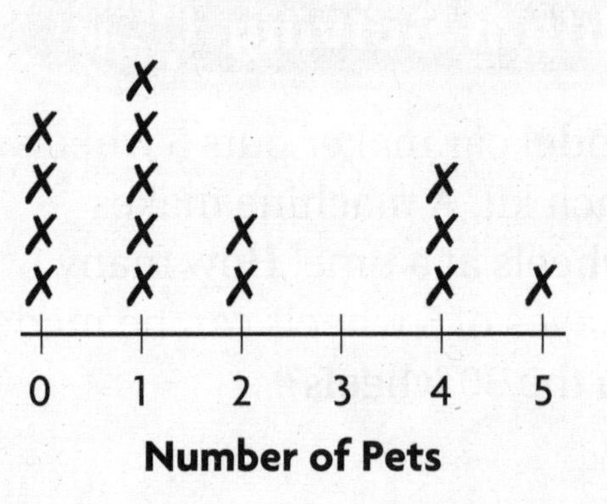

FOR MORE PRACTICE GO TO THE Personal Math Trainer

Name ______________________________

Divide by 3

Essential Question What strategies can you use to divide by 3?

Common Core **Operations and Algebraic Thinking—3.OA.C.7** *Also 3.OA.A.2, 3.OA.A.3, 3.OA.A.4, 3.OA.B.6*

MATHEMATICAL PRACTICES
MP1, MP4, MP5, MP6

Hands On

Unlock the Problem Real World

For field day, 18 students have signed up for the relay race. Each relay team needs 3 students. How many teams can be made?

- What do you need to find?

____

____

- Circle the numbers you need to use.

One Way Make equal groups.

- Look at the 18 counters below.
- Circle as many groups of 3 as you can.
- Count the number of groups.

There are _____ groups of 3.

So, _____ teams can be made.

You can write $18 \div 3 =$ _____ or $3\overline{)18}$.

Math Talk MATHEMATICAL PRACTICES 1

Make Sense of Problems Suppose the question asked how many students would be on 3 equal teams. How would you model 3 equal teams? Would the quotient be the same?

Other Ways

A Count back on a number line.

- Start at 18.
- Count back by 3s as many times as you can. Complete the jumps on the number line.
- Count the number of times you jumped back 3.

ERROR Alert

Be sure to count back the same number of spaces each time you jump back on the number line.

0 1 2 3 4 5 6 7 8 9 10 11 12 13 14 15 16 17 18

You jumped back by 3 _____ times.

B Use a related multiplication fact.

Since division is the opposite of multiplication, think of a related multiplication fact to find 18 ÷ 3.

$\square \times 3 = 18$

$6 \times 3 = 18$

Think: What number completes the multiplication fact?

So, 18 ÷ 3 = _____ or $3\overline{)18}$.

- What if 24 students signed up for the relay race and there were 3 students on each team? What related multiplication fact would you use to find the number of teams?

____

Share and Show MATH BOARD

1. Circle groups of 3 to find 12 ÷ 3. _____

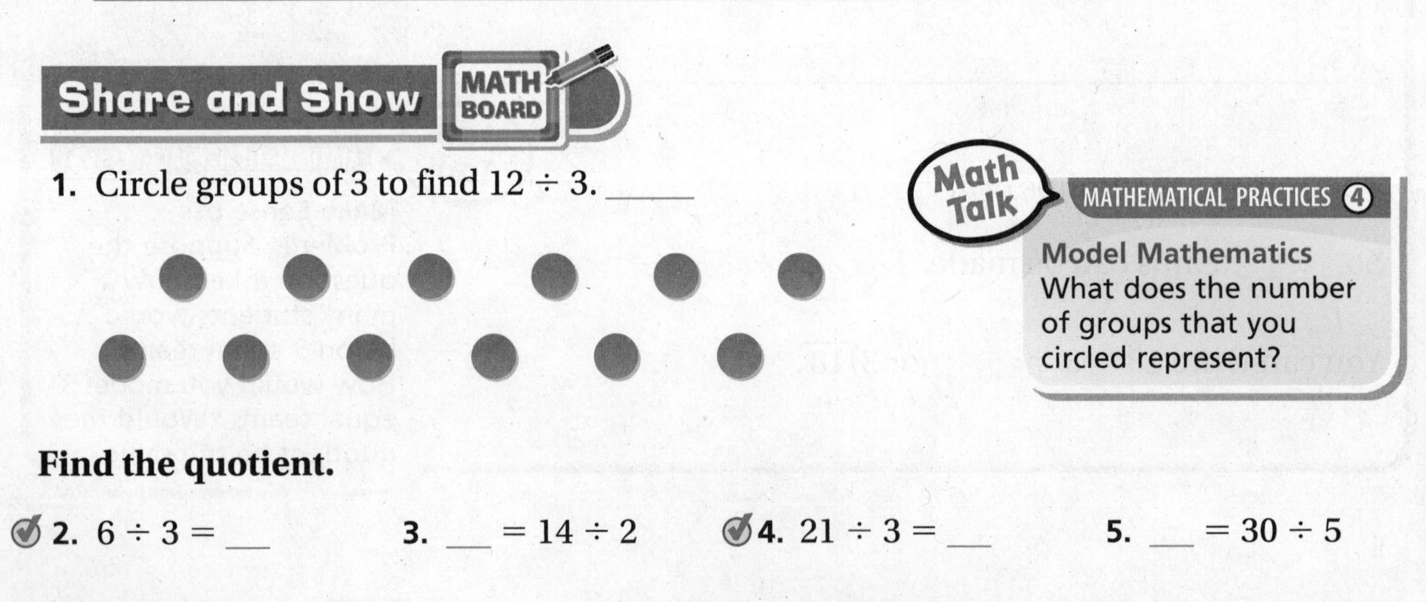

Math Talk MATHEMATICAL PRACTICES 4

Model Mathematics What does the number of groups that you circled represent?

Find the quotient.

2. 6 ÷ 3 = ___ 3. ___ = 14 ÷ 2 4. 21 ÷ 3 = ___ 5. ___ = 30 ÷ 5

Name ______________________

On Your Own

Practice: Copy and Solve **Find the quotient. Draw a quick picture to help.**

6. $9 \div 3$ **7.** $10 \div 5$ **8.** $18 \div 2$ **9.** $24 \div 3$

Find the quotient.

10. ___ $= 12 \div 2$ **11.** $40 \div 5 =$ ___ **12.** $60 \div 10 =$ ___ **13.** ___ $= 20 \div 10$

14. $3\overline{)15}$ **15.** $2\overline{)4}$ **16.** $5\overline{)20}$ **17.** $3\overline{)18}$

MATHEMATICAL PRACTICE 2 **Use Reasoning** **Algebra** **Write +, −, ×, or ÷.**

18. $25 \bigcirc 5 = 10 \div 2$

19. $3 \times 3 = 6 \bigcirc 3$

20. $16 \bigcirc 2 = 24 - 16$

21. $13 + 19 = 8 \bigcirc 4$

22. $14 \bigcirc 2 = 6 \times 2$

23. $21 \div 3 = 5 \bigcirc 2$

24. Jem pastes 21 photos and 15 postcards in a scrap album. She puts 3 on each page. How many pages does Jem fill in the scrap album?

____

25. GO DEEPER Sue plants 18 pink flowers and 9 yellow flowers in flowerpots. She plants 3 flowers in each flowerpot. How many flowerpots does Sue use?

____

26. GO DEEPER Blaine makes an array of 12 red squares and 18 blue squares. She puts 3 squares in each row. How many rows does Blaine have in the array?

____

Problem Solving • Applications Real World

Use the table for 27–28.

Field Day Events

Activity	Number of Students
Relay race	25
Beanbag toss	18
Jump-rope race	27

27. GO DEEPER There are 5 equal teams in the relay race. How many students are on each team? Write a division equation that shows the number of students on each team.

____

28. THINK SMARTER Students doing the jump-rope race and the beanbag toss compete in teams of 3. How many more teams participate in the jump-rope race than in the beanbag toss? **Explain** how you know.

WRITE Math
Show Your Work

____

____

29. MATHEMATICAL PRACTICE 1 **Make Sense of Problems** Michael puts 21 sports cards into stacks of 3. The answer is 7 stacks. What's the question?

____

____

30. THINK SMARTER Jorge made $24 selling water at a baseball game. He wants to know how many bottles of water he sold. Jorge used this number line to help him.

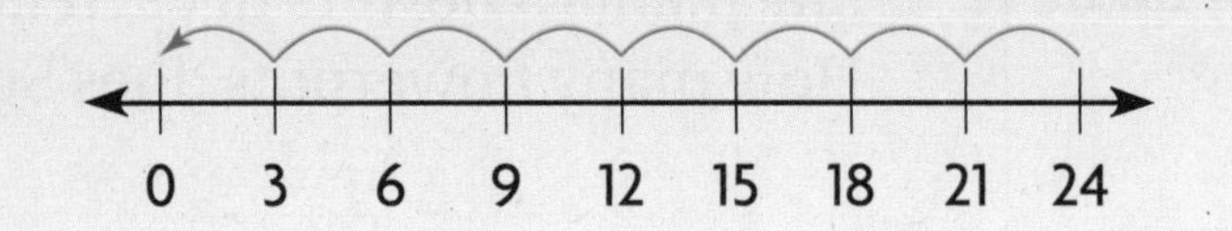

Write the division equation that the number line represents.

_____ ÷ _____ = _____

Name ____________________

Divide by 3

Common Core **COMMON CORE STANDARD—3.OA.C.7**
Multiply and divide within 100.

Find the quotient. Draw a quick picture to help.

1. 12 ÷ 3 = 4

2. 24 ÷ 3 = ____

3. ____ = 6 ÷ 3

4. 40 ÷ 5 = ____

Find the quotient.

5. ____ = 15 ÷ 3

6. ____ = 21 ÷ 3

7. 16 ÷ 2 = ____

8. 27 ÷ 3 = ____

9. 0 ÷ 3 = ____

10. 9 ÷ 3 = ____

11. ____ = 30 ÷ 3

12. ____ = 12 ÷ 4

13. $3\overline{)12}$

14. $3\overline{)15}$

15. $3\overline{)24}$

16. $3\overline{)9}$

Problem Solving Real World

17. The principal at Miller Street School has 12 packs of new pencils. She will give 3 packs to each third-grade class. How many third-grade classes are there?

____

18. Mike has $21 to spend at the mall. He spends all of his money on bracelets for his sisters. Bracelets cost $3 each. How many bracelets does he buy?

____

19. WRITE Math Explain how to divide an amount by 3.

____

____

Lesson Check (3.OA.C.7)

1. There are 18 counters divided equally among 3 groups. How many counters are in each group?

____

2. Josh has 27 signed baseballs. He places the baseballs equally on 3 shelves. How many baseballs are on each shelf?

____

Spiral Review (3.OA.A.1, 3.OA.B.5, 3.OA.B.6, 3.MD.B.4)

3. Each bicycle has 2 wheels. How many wheels do 8 bicycles have?

____

4. How many students watch less than 3 hours of TV a day?

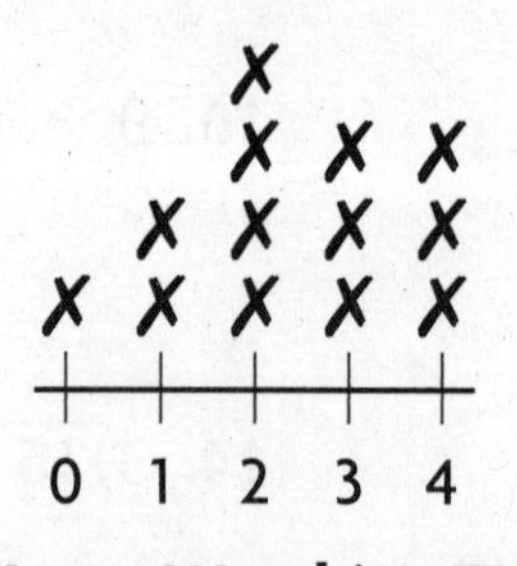

____

5. Complete the number sentence to show an example of the Distributive Property.

$3 \times 6 =$

____

____

____

____

6. What unknown number completes the equations?

$3 \times \blacksquare = 21 \qquad 21 \div 3 = \blacksquare$

____

____

____

____

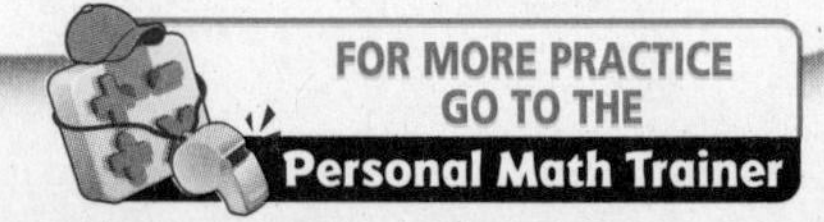

Name ______________________________

Divide by 4

Essential Question What strategies can you use to divide by 4?

Common Core **Operations and Algebraic Thinking—3.OA.C.7** *Also 3.OA.A.2, 3.OA.A.3, 3.OA.A.4, 3.OA.B.5, 3.OA.B.6*

MATHEMATICAL PRACTICES
MP3, MP4, MP7, MP8

Unlock the Problem (Real World)

A tree farmer plants 12 red maple trees in 4 equal rows. How many trees are in each row?

- What strategy could you use to solve the problem?

____

One Way Make an array.

- Look at the array.
- Continue the array by drawing 1 tile in each of the 4 rows until all 12 tiles are drawn.
- Count the number of tiles in each row.

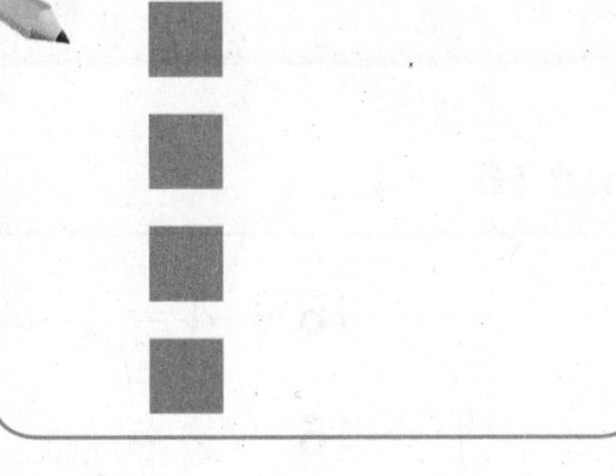

There are _____ tiles in each row.

So, there are _____ trees in each row.

Write: _____ ÷ _____ = _____ or $4\overline{)12}$

Read: Twelve divided by four equals three.

Other Ways

A Make equal groups.

- Draw 1 counter in each group.
- Continue drawing 1 counter at a time until all 12 counters are drawn.

There are _____ counters in each group.

Math Talk MATHEMATICAL PRACTICES 6

Compare How is making an array to solve the problem like making equal groups?

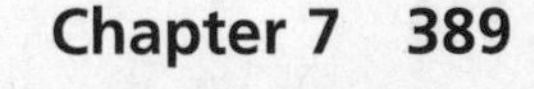

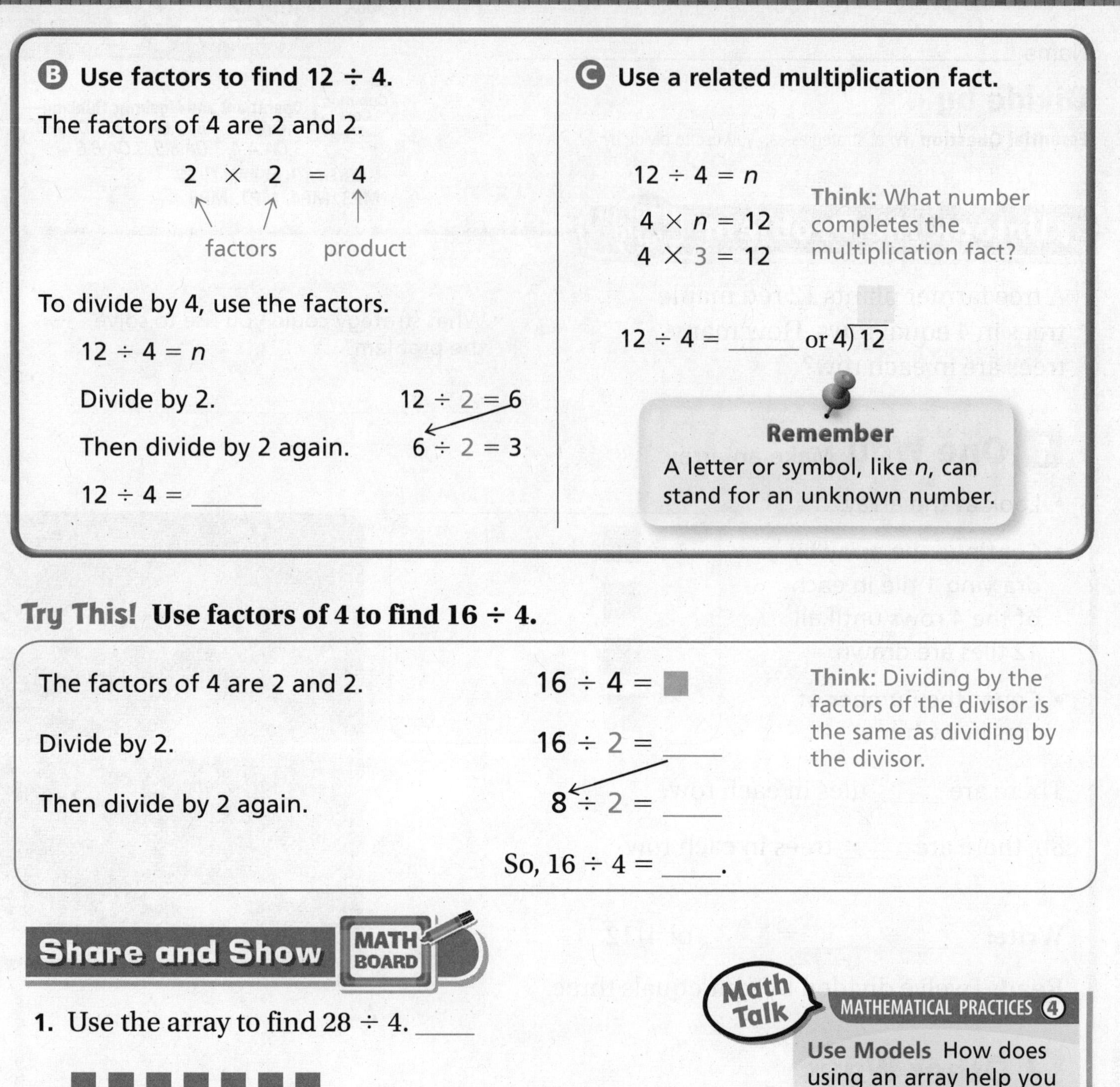

B Use factors to find $12 \div 4$.

The factors of 4 are 2 and 2.

$2 \times 2 = 4$

factors product

To divide by 4, use the factors.

$12 \div 4 = n$

Divide by 2. $12 \div 2 = 6$

Then divide by 2 again. $6 \div 2 = 3$

$12 \div 4 =$ _____

C Use a related multiplication fact.

$12 \div 4 = n$

$4 \times n = 12$

$4 \times 3 = 12$

Think: What number completes the multiplication fact?

$12 \div 4 =$ _____ or $4\overline{)12}$

Remember

A letter or symbol, like *n*, can stand for an unknown number.

Try This! Use factors of 4 to find $16 \div 4$.

The factors of 4 are 2 and 2. $16 \div 4 = ■$

Divide by 2. $16 \div 2 =$ _____

Then divide by 2 again. $8 \div 2 =$ _____

Think: Dividing by the factors of the divisor is the same as dividing by the divisor.

So, $16 \div 4 =$ _____.

Share and Show MATH BOARD

1. Use the array to find $28 \div 4$. _____

Math Talk MATHEMATICAL PRACTICES 4

Use Models How does using an array help you to find a quotient?

Find the quotient.

2. ___ $= 21 \div 3$
3. $8 \div 4 =$ ___
4. ___ $= 40 \div 5$
5. $24 \div 4 =$ ___

Find the unknown number.

6. $20 \div 4 = a$ $a =$ ___
7. $12 \div 2 = p$ $p =$ ___
8. $27 \div 3 = ▲$ $▲ =$ ___
9. $12 \div 4 = t$ $t =$ ___

Name ______________________

On Your Own

Practice: Copy and Solve **Draw tiles to make an array. Find the quotient.**

10. $30 \div 10$ **11.** $15 \div 5$ **12.** $40 \div 4$ **13.** $16 \div 2$

Find the quotient.

14. $12 \div 3 =$ ___ **15.** $20 \div 4 =$ ___ **16.** $4\overline{)16}$ **17.** $5\overline{)25}$

Find the unknown number.

18. $45 \div 5 = b$ $b =$ ___

19. $20 \div 10 = e$ $e =$ ___

20. $8 \div 2 = ■$ $■ =$ ___

21. $24 \div 3 = h$ $h =$ ___

Algebra **Complete the table.**

22.

÷	9	12	15	18
3				

23.

÷	20	24	28	32
4				

MATHEMATICAL PRACTICE 2 **Use Reasoning** **Algebra** **Find the unknown number.**

24. $14 \div$ ___ $= 7$ **25.** $30 \div$ ___ $= 6$ **26.** $8 \div$ ___ $= 2$ **27.** $24 \div$ ___ $= 8$

28. $36 \div$ ___ $= 9$ **29.** $40 \div$ ___ $= 4$ **30.** $3 \div$ ___ $= 1$ **31.** $35 \div$ ___ $= 7$

32. Mr. Benz arranges 24 music stands in class. He puts the stands in 4 equal rows. How many music stands are in each row?

____

33. GO DEEPER Monty has 16 toy cars in 4 equal groups and 24 toy boats in 3 equal groups. How many more toy boats are in each group than toy cars?

____

34. GO DEEPER Mia puts 15 animal stickers in 3 equal rows in her sticker book. She puts 28 flower stickers in 4 equal rows. How many more flower stickers than animal stickers are in each row?

____

Problem Solving • Applications Real World

Use the table for 35–36.

35. GO DEEPER Douglas planted the birch trees in 4 equal rows. Then he added 2 more birch trees to each row. How many birch trees did he plant in each row?

____

36. THINK SMARTER Mrs. Banks planted the oak trees in 4 equal rows. Mr. Webb planted the dogwood trees in 3 equal rows. Who planted more trees in each row? How many more? Explain how you know.

Math on the Spot

Trees Planted	
Type	**Number Planted**
Dogwood	24
Oak	28
Birch	16

____

____

____

WRITE Math
Show Your Work

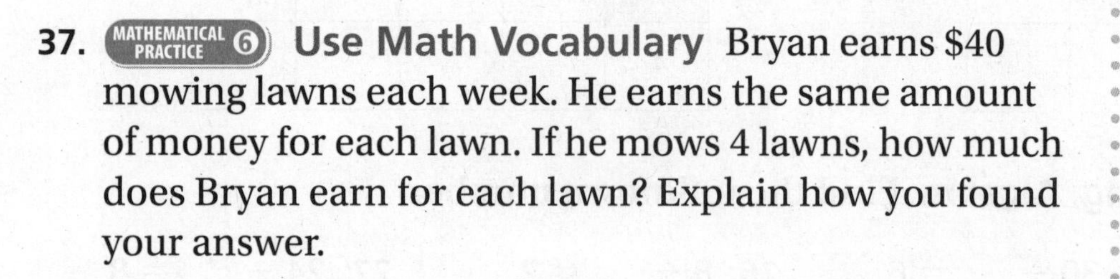

37. MATHEMATICAL PRACTICE 6 **Use Math Vocabulary** Bryan earns $40 mowing lawns each week. He earns the same amount of money for each lawn. If he mows 4 lawns, how much does Bryan earn for each lawn? Explain how you found your answer.

____

____

____

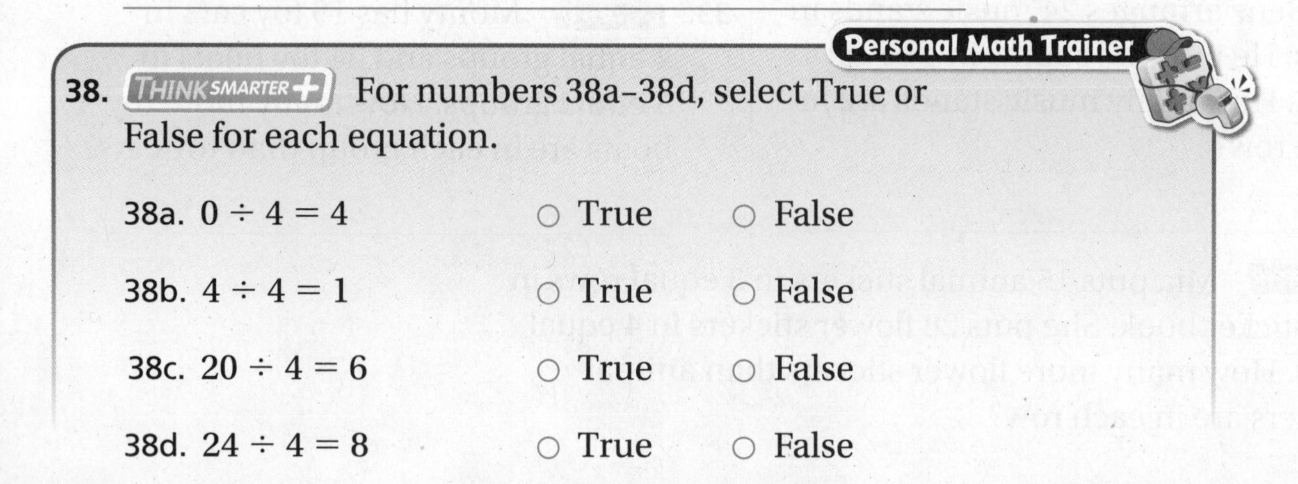

Personal Math Trainer

38. THINK SMARTER + For numbers 38a–38d, select True or False for each equation.

38a. $0 \div 4 = 4$ ○ True ○ False

38b. $4 \div 4 = 1$ ○ True ○ False

38c. $20 \div 4 = 6$ ○ True ○ False

38d. $24 \div 4 = 8$ ○ True ○ False

Practice and Homework
Lesson 7.5

Name ______________________

Divide by 4

Common Core **COMMON CORE STANDARD—3.OA.C.7**
Multiply and divide within 100.

Draw tiles to make an array. Find the quotient.

1. 4 = 16 ÷ 4
2. 20 ÷ 4 = ____
3. 12 ÷ 4 = ____
4. 10 ÷ 2 = ____

Find the quotient.

5. 24 ÷ 3 = ____
6. ____ = 8 ÷ 2
7. 32 ÷ 4 = ____
8. ____ = 28 ÷ 4
9. $4\overline{)36}$
10. $4\overline{)8}$
11. $4\overline{)24}$
12. $3\overline{)30}$

Find the unknown number.

13. 20 ÷ 5 = a

a = ____

14. 32 ÷ 4 = p

p = ____

15. 40 ÷ 10 = ■

■ = ____

16. 18 ÷ 3 = x

x = ____

Problem Solving Real World

17. Ms. Higgins has 28 students in her gym class. She puts them in 4 equal groups. How many students are in each group?

____

18. Andy has 36 CDs. He buys a case that holds 4 CDs in each section. How many sections can he fill?

____

19. WRITE Math Write and solve a word problem that involves dividing by 4.

____

____

Lesson Check (3.OA.C.7)

1. Darion picks 16 grapefruits off a tree in his backyard. He puts 4 grapefruits in each bag. How many bags does he need?

2. Tori has a bag of 32 markers to share equally among 3 friends and herself. How many markers will Tori and each of her friends get?

Spiral Review (3.OA.A.2, 3.OA.B.5, 3.OA.C.7, 3.OA.D.9)

3. Find the product.

3×7

4. Describe a pattern below.

8, 12, 16, 20, 24, 28

5. Use the Commutative Property of Multiplication to write a related number sentence.

$4 \times 5 = 20$

6. Jasmine has 18 model horses. She places the model horses equally on 3 shelves. How many model horses are on each shelf?

FOR MORE PRACTICE GO TO THE **Personal Math Trainer**

Name ____

Divide by 6

Essential Question What strategies can you use to divide by 6?

Common Core **Operations and Algebraic Thinking—3.OA.C.7** *Also 3.OA.A.2, 3.OA.A.3, 3.OA.A.4, 3.OA.B.5, 3.OA.B.6*

MATHEMATICAL PRACTICES
MP2, MP4, MP5, MP6

Unlock the Problem (Real World)

Hands On

Ms. Sing needs to buy 24 juice boxes for the class picnic. Juice boxes come in packs of 6. How many packs does Ms. Sing need to buy?

- Circle the number that tells you how many juice boxes come in a pack.
- How can you use the information to solve the problem?

____

____

____

One Way Make equal groups.

- Draw 24 counters.
- Circle as many groups of 6 as you can.
- Count the number of groups.

There are _____ groups of 6.

So, Ms. Sing needs to buy _____ packs of juice boxes.

You can write ____ ÷ ____ = ____ or $6\overline{)24}$.

Math Talk MATHEMATICAL PRACTICES 1

Make Sense of Problems If you divided the 24 counters into groups of 4, how many groups would there be?

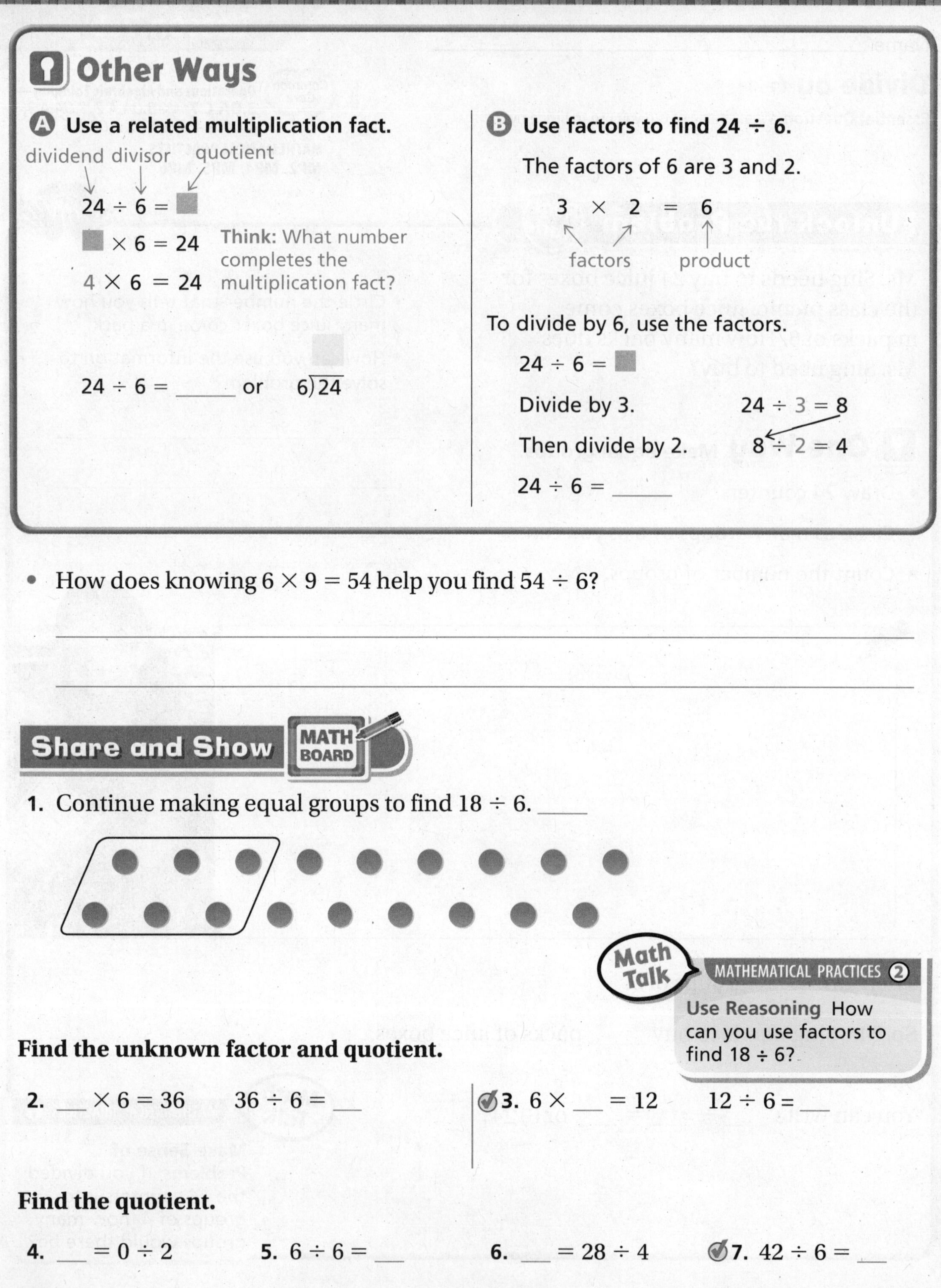

Other Ways

A Use a related multiplication fact.

dividend divisor quotient

$24 \div 6 = ■$

$■ \times 6 = 24$

$4 \times 6 = 24$

Think: What number completes the multiplication fact?

$24 \div 6 =$ ____ or $6\overline{)24}$

B Use factors to find 24 ÷ 6.

The factors of 6 are 3 and 2.

$3 \times 2 = 6$

factors product

To divide by 6, use the factors.

$24 \div 6 = ■$

Divide by 3. $24 \div 3 = 8$

Then divide by 2. $8 \div 2 = 4$

$24 \div 6 =$ ____

- How does knowing $6 \times 9 = 54$ help you find $54 \div 6$?

Share and Show MATH BOARD

1. Continue making equal groups to find $18 \div 6$. ____

Math Talk — MATHEMATICAL PRACTICES 2

Use Reasoning How can you use factors to find $18 \div 6$?

Find the unknown factor and quotient.

2. ____ $\times 6 = 36$ $36 \div 6 =$ ____

3. $6 \times$ ____ $= 12$ $12 \div 6 =$ ____

Find the quotient.

4. ____ $= 0 \div 2$

5. $6 \div 6 =$ ____

6. ____ $= 28 \div 4$

7. $42 \div 6 =$ ____

Name ___________________________

On Your Own

Find the unknown factor and quotient.

8. $6 \times$ ____ $= 30$ $30 \div 6 =$ ____

9. ____ $\times 6 = 48$ $48 \div 6 =$ ____

Find the quotient.

10. $12 \div 6 =$ ____

11. ____ $= 6 \div 1$

12. $6\overline{)6}$

13. $2\overline{)10}$

Find the unknown number.

14. $24 \div 6 = n$

$n =$ ____

15. $40 \div 5 =$ ▲

▲ $=$ ____

16. $60 \div 10 = m$

$m =$ ____

17. $18 \div 6 =$ ■

■ $=$ ____

MATHEMATICAL PRACTICE 2 **Use Reasoning** **Algebra** **Find the unknown number.**

18. $20 \div$ ____ $= 4$

19. $24 \div$ ____ $= 8$

20. $16 \div$ ____ $= 4$

21. $3 \div$ ____ $= 3$

22. $42 \div$ ____ $= 7$

23. $30 \div$ ____ $= 10$

24. $10 \div$ ____ $= 2$

25. $32 \div$ ____ $= 4$

26. Mr. Brooks has 36 students in his gym class. He makes 6 teams. If each team has the same number of students, how many students are on each team?

____

27. GO DEEPER Sandy bakes 18 pies. She keeps 2 of the pies. She sells the rest of the pies to 4 people at a bake sale. If each person buys the same number of pies, how many pies does Sandy sell to each person?

____

28. THINK SMARTER Derek has 2 boxes of fruit snacks. There are 12 fruit snacks in each box. If he eats 6 fruit snacks each day, how many days will the fruit snacks last? Explain.

____

____

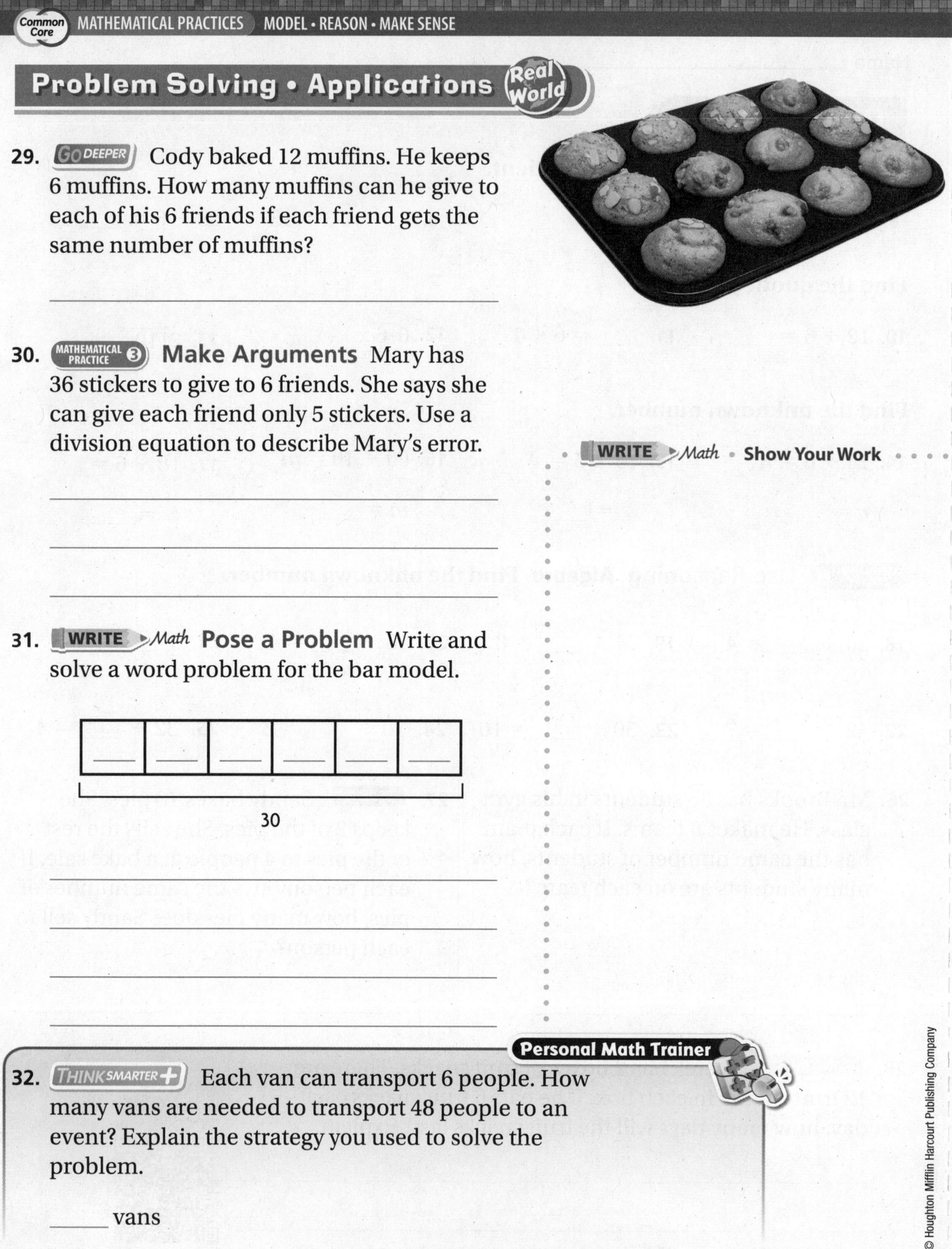

Problem Solving • Applications Real World

29. GO DEEPER Cody baked 12 muffins. He keeps 6 muffins. How many muffins can he give to each of his 6 friends if each friend gets the same number of muffins?

____

30. MATHEMATICAL PRACTICE 3 **Make Arguments** Mary has 36 stickers to give to 6 friends. She says she can give each friend only 5 stickers. Use a division equation to describe Mary's error.

____

____

____

31. WRITE Math **Pose a Problem** Write and solve a word problem for the bar model.

___	___	___	___	___	___

30

____

____

____

____

WRITE Math • Show Your Work

Personal Math Trainer

32. THINK SMARTER+ Each van can transport 6 people. How many vans are needed to transport 48 people to an event? Explain the strategy you used to solve the problem.

_____ vans

Name ______________________

Divide by 6

Common Core **COMMON CORE STANDARD—3.OA.C.7**
Multiply and divide within 100.

Find the unknown factor and quotient.

1. $6 \times$ __7__ $= 42$ $42 \div 6 =$ __7__
2. $6 \times$ ____ $= 18$ $18 \div 6 =$ ____
3. $4 \times$ ____ $= 24$ $24 \div 4 =$ ____
4. $6 \times$ ____ $= 54$ $54 \div 6 =$ ____

Find the quotient.

5. ____ $= 24 \div 6$
6. $48 \div 6 =$ ____
7. ____ $= 6 \div 6$
8. $12 \div 6 =$ ____
9. $6\overline{)36}$
10. $6\overline{)54}$
11. $6\overline{)30}$
12. $1\overline{)6}$

Find the unknown number.

13. $p = 42 \div 6$

    $p =$ ____
14. $18 \div 3 = q$

    $q =$ ____
15. $r = 30 \div 6$

    $r =$ ____
16. $60 \div 6 = s$

    $s =$ ____

Problem Solving Real World

17. Lucas has 36 pages of a book left to read. If he reads 6 pages a day, how many days will it take Lucas to finish the book?

    ____
18. Juan has $24 to spend at the bookstore. If books cost $6 each, how many books can he buy?

    ____
19. WRITE Math Which strategy would you use to divide $36 \div 6$? Explain why you chose that strategy.

    ____

    ____

Lesson Check (3.OA.C.7)

1. Ella earned $54 last week babysitting. She earns $6 an hour. How many hours did Ella babysit last week?

2. What is the unknown factor and quotient?

   $6 \times \blacksquare = 42 \qquad 42 \div 6 = \blacksquare$

Spiral Review (3.OA.A.1, 3.OA.A.2, 3.OA.C.7, 3.OA.D.8)

3. Coach Clarke has 48 students in his P.E. class. He places the students in teams of 6 for an activity. How many teams can Coach Clarke make?

4. Each month for 7 months, Eva reads 3 books. How many more books does she need to read before she has read 30 books?

5. Each cow has 4 legs. How many legs do 5 cows have?

6. Find the product.

   3×9

Name ________________________________

Concepts and Skills

1. **Explain** how to find $20 \div 4$ by making an array. (3.OA.A.3)

____

____

____

2. **Explain** how to find $30 \div 6$ by making equal groups. (3.OA.A.3)

____

____

____

Find the unknown factor and quotient. (3.OA.C.7)

3. $10 \times$ ___ $= 50$ ___ $= 50 \div 10$

4. $2 \times$ ___ $= 16$ ___ $= 16 \div 2$

5. $2 \times$ ___ $= 20$ ___ $= 20 \div 2$

6. $5 \times$ ___ $= 20$ ___ $= 20 \div 5$

Find the quotient. (3.OA.A.3, 3.OA.C.7)

7. ___ $= 6 \div 6$

8. $21 \div 3 =$ ___

9. ___ $= 0 \div 3$

10. $36 \div 4 =$ ___

11. $5\overline{)35}$

12. $4\overline{)24}$

13. $6\overline{)54}$

14. $3\overline{)9}$

15. Carter has 18 new books. He plans to read 3 of them each week. How many weeks will it take Carter to read all of his new books? (3.OA.C.7)

____

16. GO DEEPER Gabriella made 5 waffles for breakfast. She has 25 strawberries and 15 blueberries to put on top of the waffles. She will put an equal number of berries on each waffle. How many berries will Gabriella put on each waffle? (3.OA.A.3)

____

17. There are 60 people at the fair waiting in line for a ride. Each car in the ride can hold 10 people. Write an equation that could be used to find the number of cars needed to hold all 60 people. (3.OA.C.7)

____

18. Alyssa has 4 cupcakes. She gives 2 cupcakes to each of her cousins. How many cousins does Alyssa have? (3.OA.A.3)

____

Name ___________________________

Divide by 7

Essential Question What strategies can you use to divide by 7?

Common Core **Operations and Algebraic Thinking—3.OA.C.7** *Also 3.OA.A.2, 3.OA.A.3, 3.OA.A.4, 3.OA.B.6*

MATHEMATICAL PRACTICES
MP2, MP4, MP6, MP8

Unlock the Problem — Real World — Hands On

Yasmin used 28 large apples to make 7 loaves of apple bread. She used the same number of apples for each loaf. How many apples did Yasmin use for each loaf?

- Do you need to find the number of equal groups or the number in each group?

____

- What label will your answer have?

____

One Way Make an array.

- Draw 1 tile in each of 7 rows.
- Continue drawing 1 tile in each of the 7 rows until all 28 tiles are drawn.
- Count the number of tiles in each row.

There are ____ tiles in each row.

So, Yasmin used ________ for each loaf.

You can write $28 \div 7 =$ ____ or $7\overline{)28}$.

Math Talk MATHEMATICAL PRACTICES 1

Make Sense of Problems Why can you use division to solve the problem?

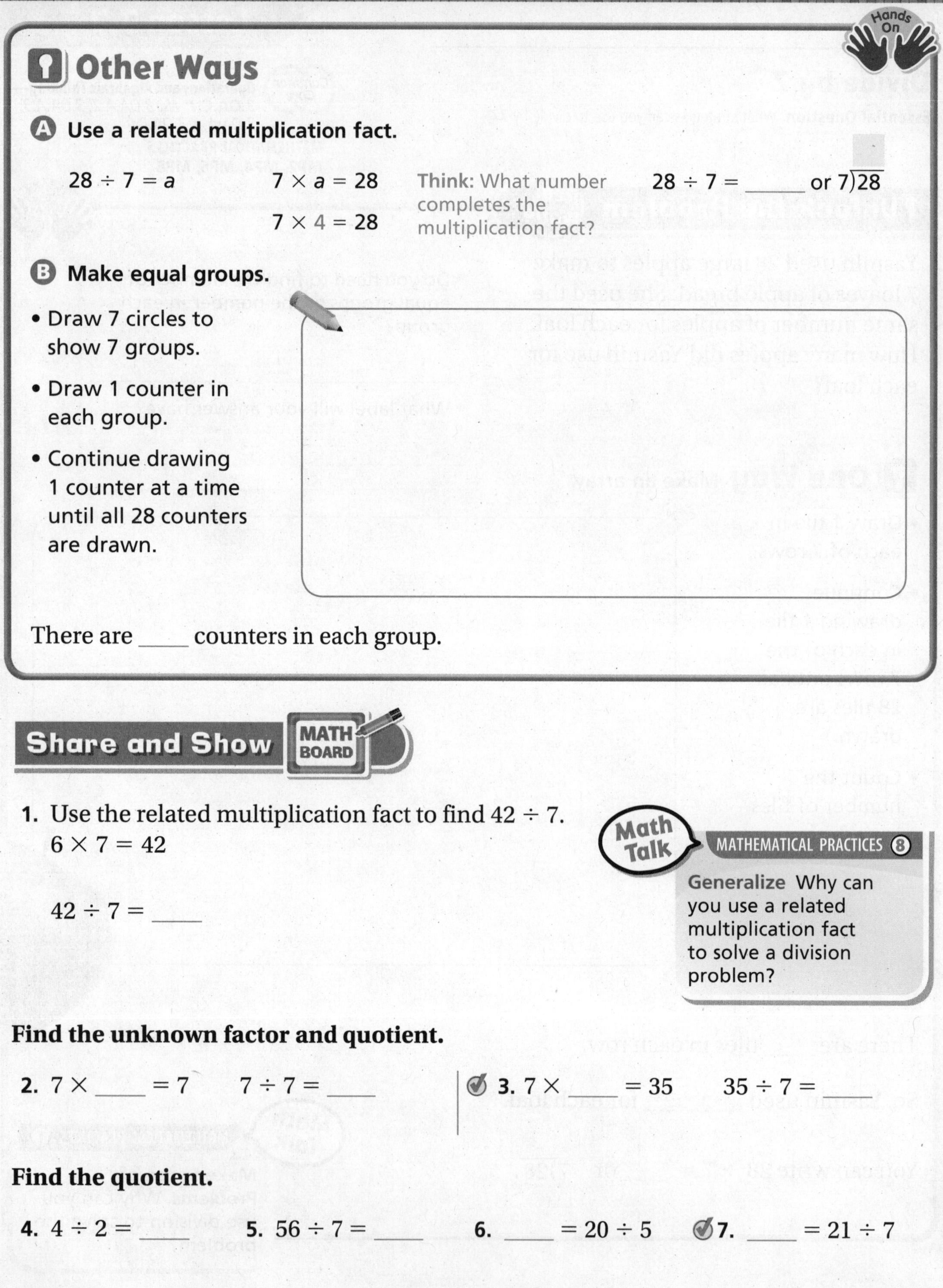

Hands On

Other Ways

A Use a related multiplication fact.

$28 \div 7 = a$ $7 \times a = 28$

$7 \times 4 = 28$

Think: What number completes the multiplication fact?

$28 \div 7 =$ ____ or $7\overline{)28}$

B Make equal groups.

- Draw 7 circles to show 7 groups.
- Draw 1 counter in each group.
- Continue drawing 1 counter at a time until all 28 counters are drawn.

There are ____ counters in each group.

Share and Show

MATH BOARD

1. Use the related multiplication fact to find $42 \div 7$.
   $6 \times 7 = 42$

   $42 \div 7 =$ ____

Math Talk MATHEMATICAL PRACTICES 8

Generalize Why can you use a related multiplication fact to solve a division problem?

Find the unknown factor and quotient.

2. $7 \times$ ____ $= 7$ $7 \div 7 =$ ____

3. $7 \times$ ____ $= 35$ $35 \div 7 =$ ____

Find the quotient.

4. $4 \div 2 =$ ____

5. $56 \div 7 =$ ____

6. ____ $= 20 \div 5$

7. ____ $= 21 \div 7$

Name ______________________

On Your Own

Find the unknown factor and quotient.

8. $3 \times \_\_\_\_ = 9$ $\_\_\_\_ = 9 \div 3$

9. $7 \times \_\_\_\_ = 49$ $49 \div 7 = \_\_\_\_$

Find the quotient.

10. $48 \div 6 = \_\_\_\_$

11. $7 \div 1 = \_\_\_\_$

12. $7\overline{)21}$

13. $2\overline{)8}$

Find the unknown number.

14. $60 \div 10 = ■$ ■ $= \_\_\_\_$

15. $70 \div 7 = k$ $k = \_\_\_\_$

16. $m = 63 \div 9$ $m = \_\_\_\_$

17. $r = 12 \div 6$ $r = \_\_\_\_$

MATHEMATICAL PRACTICE 6 **Make Connections** **Algebra** **Complete the table.**

18.

÷	18	30	24	36
6				

19.

÷	56	42	49	35
7				

20. Clare bought 35 peaches to make peach jam. She used 7 peaches for each jar of jam. How many jars did Clare make?

____

21. There are 49 jars of peach salsa packed into 7 gift boxes. If each box has the same number of jars of salsa, how many jars are in each box?

____

22. GO DEEPER There are 31 girls and 25 boys in the marching band. When the band marches, they are in 7 rows. How many people are in each row?

____

23. GO DEEPER Ed has 42 red beads and 28 blue beads. He uses an equal number of all the beads to decorate 7 clay sculptures. How many beads are on each sculpture?

____

Unlock the Problem Real World

24. THINK SMARTER Gavin sold 21 bagels to 7 different people. Each person bought the same number of bagels. How many bagels did Gavin sell to each person?

Math on the Spot

a. What do you need to find? ______________________________

____

b. How can you use a bar model to help you decide which operation to use to solve the problem? ______________________________

____

____

c. Complete the bar model to help you find the number of bagels Gavin sold to each person.

___	___	___	___	___	___	___

21 bagels

d. What is another way you could have solved the problem?

____

____

e. Complete the sentences.

Gavin sold ____ bagels to ____ different people.

Each person bought the same number of __________.

So, Gavin sold _____ bagels to each person.

25. GO DEEPER There are 35 plain bagels and 42 wheat bagels on 7 shelves in the bakery. Each shelf has the same number of plain bagels and the same number of wheat bagels. How many bagels are on each shelf?

____

26. THINK SMARTER Write the correct symbol that makes the equations true.

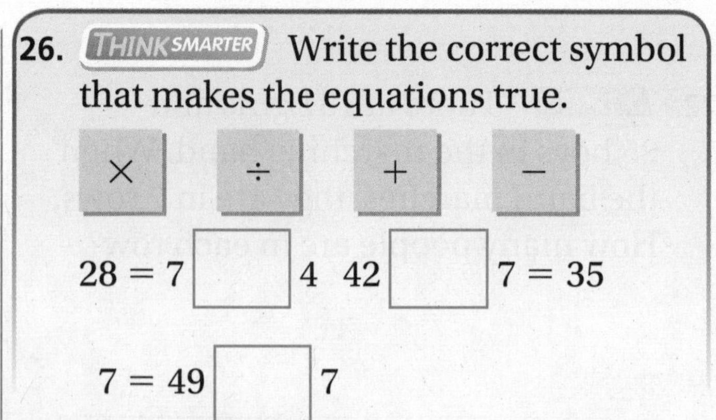

× ÷ + −

$28 = 7$ ☐ 4 42 ☐ $7 = 35$

$7 = 49$ ☐ 7

Name ______________________

Divide by 7

Common Core COMMON CORE STANDARD—3.OA.C.7
Multiply and divide within 100.

Find the unknown factor and quotient.

1. $7 \times$ __6__ $= 42$ $42 \div 7 =$ __6__
2. $7 \times$ ____ $= 35$ $35 \div 7 =$ ____
3. $7 \times$ ____ $= 7$ $7 \div 7 =$ ____
4. $5 \times$ ____ $= 20$ $20 \div 5 =$ ____

Find the quotient.

5. $7\overline{)21}$
6. $7\overline{)14}$
7. $6\overline{)48}$
8. $7\overline{)63}$
9. ____ $= 35 \div 7$
10. $0 \div 7 =$ ____
11. ____ $= 56 \div 7$
12. $32 \div 8 =$ ____

Find the unknown number.

13. $56 \div 7 = e$

 $e =$ ______

14. $k = 32 \div 4$

 $k =$ ______

15. $g = 49 \div 7$

 $g =$ ______

16. $28 \div 7 = s$

 $s =$ ______

17. Twenty-eight players sign up for basketball. The coach puts 7 players on each team. How many teams are there?

 ____

18. Roberto read 42 books over 7 months. He read the same number of books each month. How many books did Roberto read each month?

 ____

19. WRITE *Math* Describe how to find the number of weeks equal to 56 days.

 ____

 ____

Lesson Check (3.OA.C.7)

1. Elliot earned $49 last month walking his neighbor's dog. He earns $7 each time he walks the dog. How many times did Elliot walk his neighbor's dog last month?

2. What is the unknown factor and quotient?

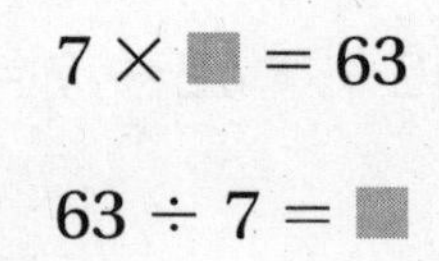

$7 \times ■ = 63$

$63 \div 7 = ■$

Spiral Review (3.OA.A.3, 3.OA.B.5, 3.OA.B.6, 3.OA.C.7)

3. Maria puts 6 strawberries in each smoothie she makes. She makes 3 smoothies. Altogether, how many strawberries does Maria use in the smoothies?

4. Kaitlyn makes 4 bracelets. She uses 8 beads for each bracelet. How many beads does she use?

5. What is the unknown factor?

$2 \times 5 = 5 \times ■$

6. What division equation is related to the following multiplication equation?

$3 \times 4 = 12$

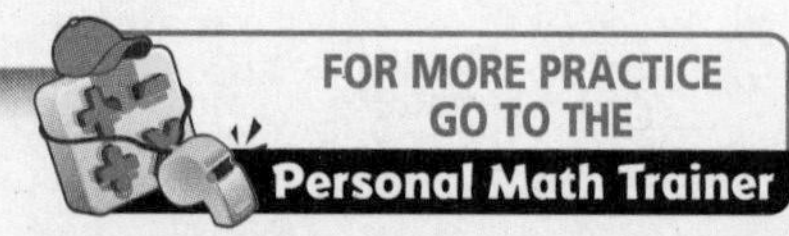

Name ____________________

Divide by 8

Essential Question What strategies can you use to divide by 8?

Common Core **Operations and Algebraic Thinking—3.OA.A.3, 3.OA.A.4** *Also 3.OA.A.2, 3.OA.B.6, 3.OA.C.7*

MATHEMATICAL PRACTICES
MP2, MP4, MP6, MP7

Unlock the Problem (Real World)

At Stephen's camping store, firewood is sold in bundles of 8 logs. He has 32 logs to put in bundles. How many bundles of firewood can he make?

- What will Stephen do with the 32 logs?

____

____

One Way Use repeated subtraction.

- Start with 32.
- Subtract 8 until you reach 0.
- Count the number of times you subtract 8.

ERROR Alert

Continue to subtract the divisor, 8, until the difference is less than 8.

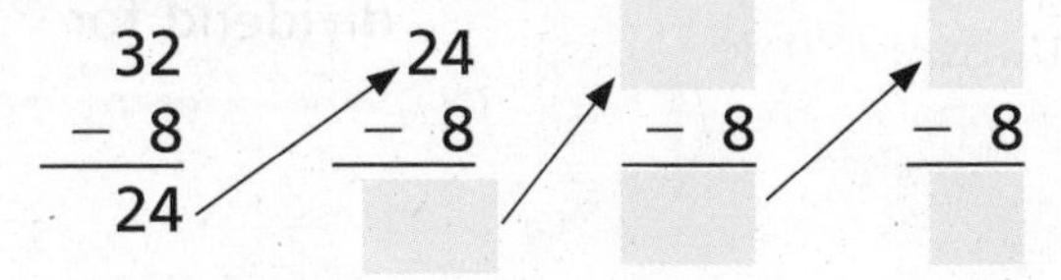

Number of times you subtract 8: 1 2 3 4

You subtracted 8 _____ times.

So, Stephen can make _____ bundles of firewood.

You can write $32 \div 8 =$ _____ or $8\overline{)32}$.

Another Way Use a related multiplication fact.

$4 \times 8 = 32$

Think: What number completes the multiplication fact?

$32 \div 8 =$ _____ or $8\overline{)32}$

Math Talk MATHEMATICAL PRACTICES 1

Make Sense of Problems How does knowing $4 \times 8 = 32$ help you find $32 \div 8$?

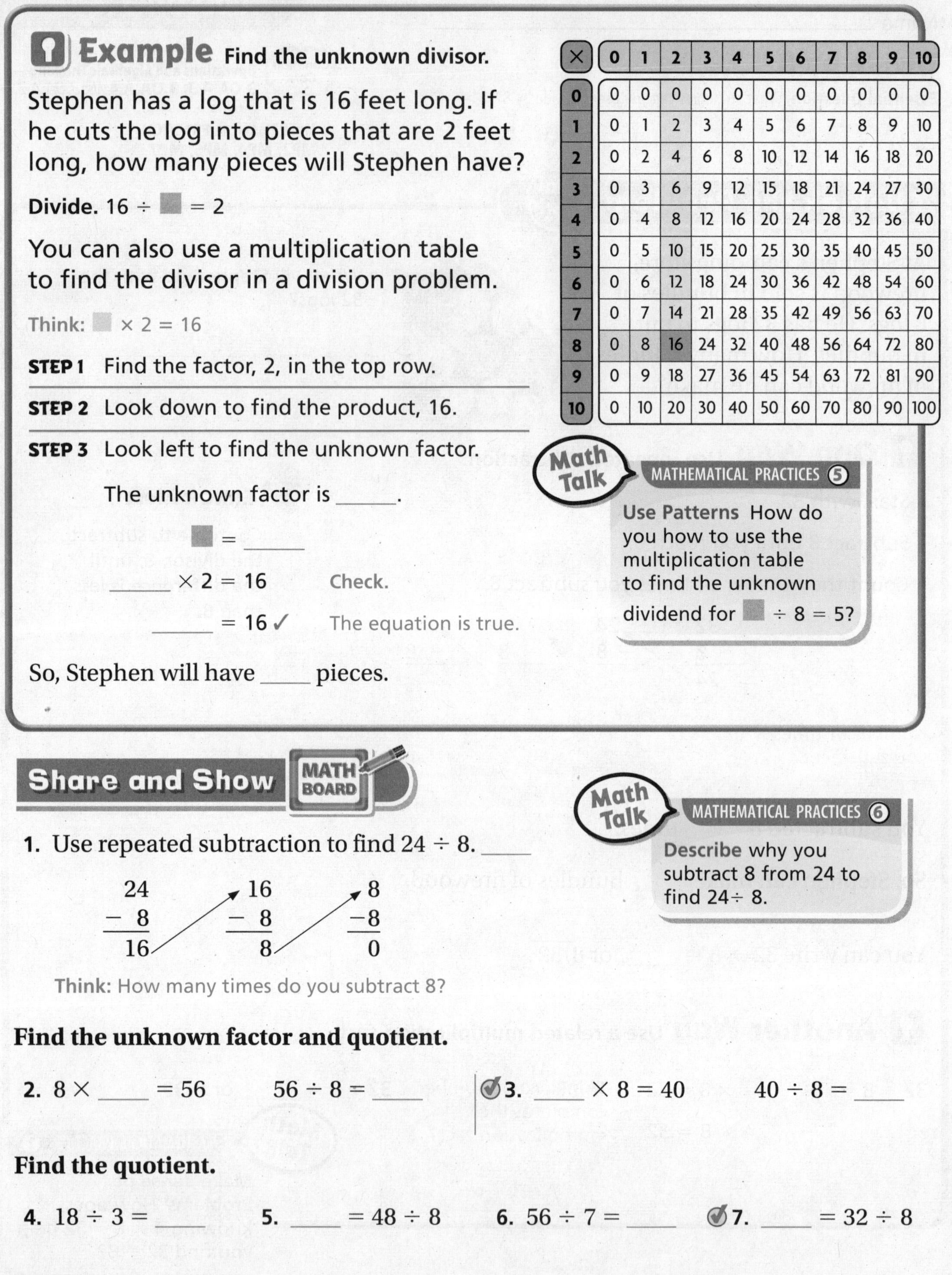

Example Find the unknown divisor.

Stephen has a log that is 16 feet long. If he cuts the log into pieces that are 2 feet long, how many pieces will Stephen have?

Divide. 16 ÷ ■ = 2

You can also use a multiplication table to find the divisor in a division problem.

Think: ■ × 2 = 16

×	0	1	2	3	4	5	6	7	8	9	10
0	0	0	0	0	0	0	0	0	0	0	0
1	0	1	2	3	4	5	6	7	8	9	10
2	0	2	4	6	8	10	12	14	16	18	20
3	0	3	6	9	12	15	18	21	24	27	30
4	0	4	8	12	16	20	24	28	32	36	40
5	0	5	10	15	20	25	30	35	40	45	50
6	0	6	12	18	24	30	36	42	48	54	60
7	0	7	14	21	28	35	42	49	56	63	70
8	0	8	16	24	32	40	48	56	64	72	80
9	0	9	18	27	36	45	54	63	72	81	90
10	0	10	20	30	40	50	60	70	80	90	100

STEP 1 Find the factor, 2, in the top row.

STEP 2 Look down to find the product, 16.

STEP 3 Look left to find the unknown factor.

The unknown factor is ____.

■ = ____

____ × 2 = 16 Check.

____ = 16 ✓ The equation is true.

So, Stephen will have ____ pieces.

Math Talk MATHEMATICAL PRACTICES 5

Use Patterns How do you how to use the multiplication table to find the unknown dividend for ■ ÷ 8 = 5?

Share and Show MATH BOARD

1. Use repeated subtraction to find 24 ÷ 8. ____

$$\begin{array}{r} 24 \\ -\ 8 \\ \hline 16 \end{array} \quad \begin{array}{r} 16 \\ -\ 8 \\ \hline 8 \end{array} \quad \begin{array}{r} 8 \\ -8 \\ \hline 0 \end{array}$$

Think: How many times do you subtract 8?

Math Talk MATHEMATICAL PRACTICES 6

Describe why you subtract 8 from 24 to find 24 ÷ 8.

Find the unknown factor and quotient.

2. 8 × ____ = 56 56 ÷ 8 = ____

✓3. ____ × 8 = 40 40 ÷ 8 = ____

Find the quotient.

4. 18 ÷ 3 = ____

5. ____ = 48 ÷ 8

6. 56 ÷ 7 = ____

✓7. ____ = 32 ÷ 8

Name ______________________

On Your Own

Find the unknown factor and quotient.

8. $6 \times$ ____ $= 18$ $18 \div 6 =$ ____

9. $8 \times$ ____ $= 72$ ____ $= 72 \div 8$

Find the quotient.

10. $28 \div 4 =$ ____

11. $42 \div 7 =$ ____

12. $8\overline{)64}$

13. $1\overline{)8}$

Find the unknown number.

14. $16 \div p = 8$

$p =$ ____

15. $t \div 8 = 2$

$t =$ ____

16. $64 \div ▲ = 8$

$▲ =$ ____

17. $m \div 8 = 10$

$m =$ ____

18. $▲ \div 2 = 10$

$▲ =$ ____

19. $40 \div ■ = 8$

$■ =$ ____

20. $25 \div k = 5$

$k =$ ____

21. $54 \div n = 9$

$n =$ ____

22. MATHEMATICAL PRACTICE 2 **Connect Symbols and Words** Write a word problem that can be solved by using one of the division facts above.

____

____

MATHEMATICAL PRACTICE 4 **Use Symbols** **Algebra** **Write +, −, ×, or ÷.**

23. $6 \times 6 = 32 \bigcirc 4$

24. $12 \bigcirc 3 = 19 - 15$

25. $40 \div 8 = 35 \bigcirc 7$

26. GO DEEPER Kyle has 4 packs of baseball cards. Each pack has 12 cards. If Kyle wants to equally divide the cards among him and his 7 friends, how many baseball cards will each person get?

____

Problem Solving • Applications (Real World)

Use the table for 27–28.

Tent Sizes

Type	Number of People
Cabin	10
Vista	8
Trail	4

27. GO DEEPER There are 32 people who plan to camp over the weekend. Describe two different ways the campers can sleep using 4 tents.

WRITE Math • Show Your Work

28. THINK SMARTER There are 36 people camping at Max's family reunion. They have cabin tents and vista tents. How many of each type of tent do they need to sleep exactly 36 people if each tent is filled? Explain.

29. Josh is dividing 64 bags of trail mix equally among 8 campers. How many bags of trail mix will each camper get?

30. THINK SMARTER Circle the unknown factor and quotient.

$8 \times$ [6 / 7 / 8] $= 48$

[6 / 7 / 8] $= 48 \div 8$

Name ______________________

Divide by 8

Common Core **COMMON CORE STANDARD—3.OA.A.3, 3.OA.A.4** *Represent and solve problems involving multiplication and division.*

Find the unknown factor and quotient.

1. $8 \times$ <u>4</u> $= 32$ $32 \div 8 =$ ____
2. $3 \times$ ____ $= 27$ $27 \div 3 =$ ____
3. $8 \times$ ____ $= 8$ $8 \div 8 =$ ____
4. $8 \times$ ____ $= 72$ $72 \div 8 =$ ____

Find the quotient.

5. ____ $= 24 \div 8$
6. $40 \div 8 =$ ____
7. ____ $= 56 \div 8$
8. $14 \div 2 =$ ____
9. $8\overline{)64}$
10. $7\overline{)28}$
11. $8\overline{)16}$
12. $8\overline{)48}$

Find the unknown number.

13. $72 \div ■ = 9$ ■ = ____
14. $25 \div ■ = 5$ ■ = ____
15. $24 \div a = 3$ $a =$ ____
16. $k \div 10 = 8$ $k =$ ____

Problem Solving Real World

17. Sixty-four students are going on a field trip. There is 1 adult for every 8 students. How many adults are there?

    ____

18. Mr. Chen spends $32 for tickets to a play. If the tickets cost $8 each, how many tickets does Mr. Chen buy?

    ____

19. WRITE Math Describe which strategy you would use to divide 48 by 8.

____

____

Lesson Check (3.OA.A.4)

1. Mrs. Wilke spends $72 on pies for the school fair. Each pie costs $8. How many pies does Mrs. Wilke buy for the school fair?

____

2. Find the unknown factor and quotient.

$8 \times ■ = 40$

$40 \div 8 = ■$

____

Spiral Review (3.OA.A.3, 3.OA.A.4, 3.OA.B.5)

3. Find the product.

$(3 \times 2) \times 5$

____

4. Use the Commutative Property of Multiplication to write a related multiplication sentence.

$9 \times 4 = 36$

____

5. Find the unknown factor.

$8 \times ■ = 32$

____

6. What multiplication sentence represents the array?

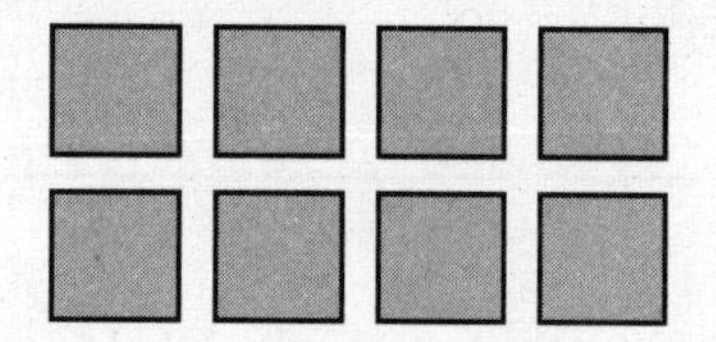

____

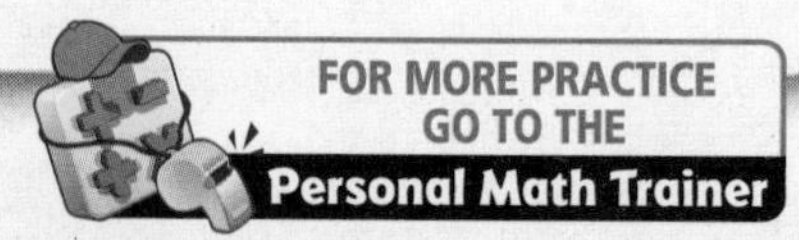

Name ______________________________

Divide by 9

Essential Question What strategies can you use to divide by 9?

Common Core **Operations and Algebraic Thinking—3.OA.C.7** *Also 3.OA.A.2, 3.OA.A.3, 3.OA.A.4, 3.OA.B.5, 3.OA.B.6*

MATHEMATICAL PRACTICES
MP2, MP4, MP6

Unlock the Problem (Real World) (Hands On)

Becket's class goes to the aquarium. The 27 students from the class are separated into 9 equal groups. How many students are in each group?

- Do you need to find the number of equal groups or the number in each group?

____

One Way Make equal groups.

- Draw 9 circles to show 9 groups.
- Draw 1 counter in each group.
- Continue drawing 1 counter at a time until all 27 counters are drawn.

There are ____ counters in each group.

So, there are ____________ in each group.

You can write $27 \div 9 =$ ____ or $9\overline{)27}$.

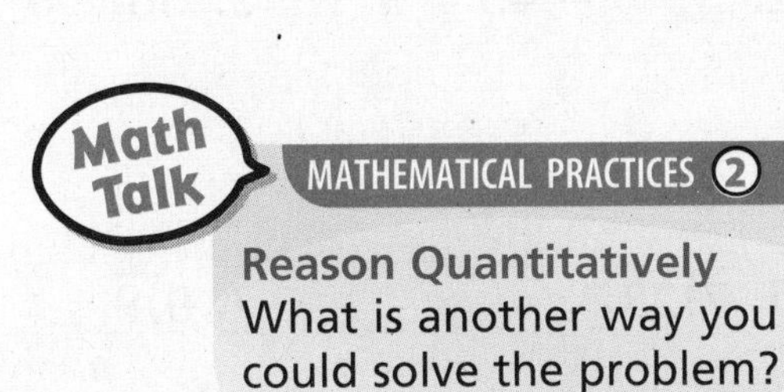

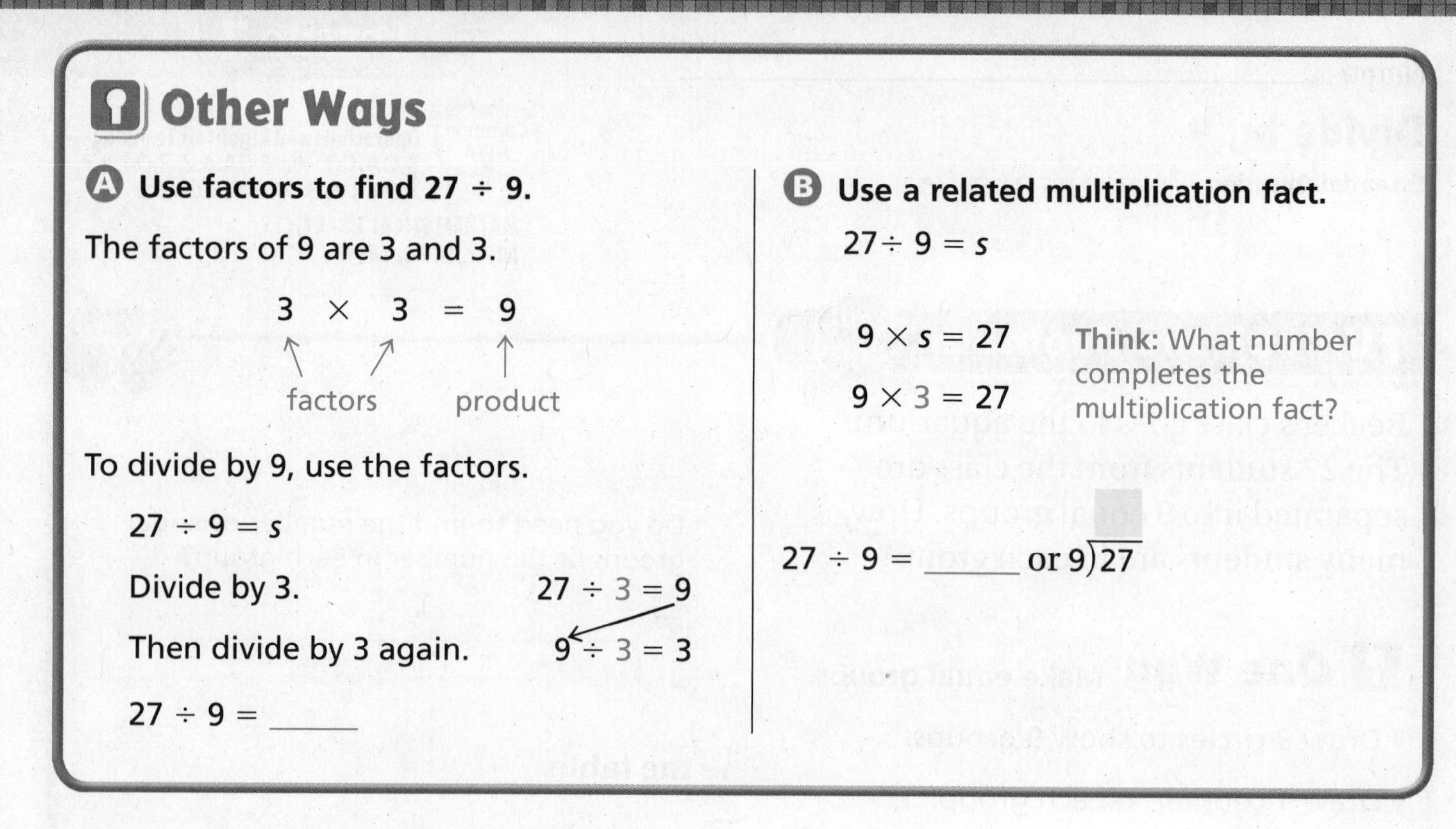

Other Ways

A **Use factors to find 27 ÷ 9.**

The factors of 9 are 3 and 3.

$3 \times 3 = 9$

To divide by 9, use the factors.

$27 \div 9 = s$

Divide by 3. $27 \div 3 = 9$

Then divide by 3 again. $9 \div 3 = 3$

$27 \div 9 =$ _____

B **Use a related multiplication fact.**

$27 \div 9 = s$

$9 \times s = 27$

$9 \times 3 = 27$

Think: What number completes the multiplication fact?

$27 \div 9 =$ _____ or $9\overline{)27}$

- What multiplication fact can you use to find 63 ÷ 9? _____________

Share and Show

1. Draw counters in the groups to find 18 ÷ 9. _____

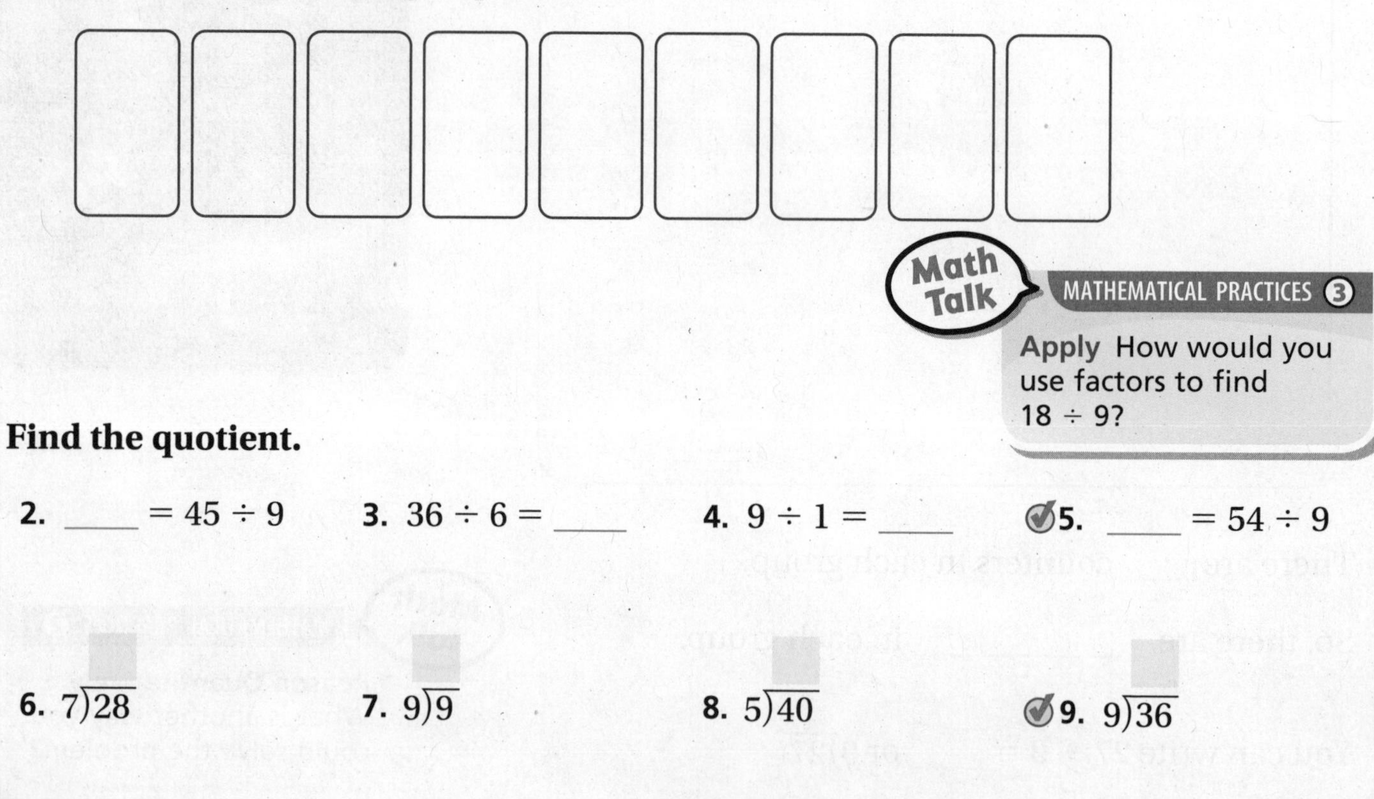

MATHEMATICAL PRACTICES 3

Apply How would you use factors to find 18 ÷ 9?

Find the quotient.

2. _____ $= 45 \div 9$ **3.** $36 \div 6 =$ _____ **4.** $9 \div 1 =$ _____ **5.** _____ $= 54 \div 9$

6. $7\overline{)28}$ **7.** $9\overline{)9}$ **8.** $5\overline{)40}$ **9.** $9\overline{)36}$

Name ______________________________

On Your Own

10. ____ $= 36 \div 4$

11. ____ $= 72 \div 9$

12. $81 \div 9 =$ ____

13. ____ $= 27 \div 9$

14. $4\overline{)12}$

15. $9\overline{)63}$

16. $2\overline{)16}$

17. $5\overline{)25}$

Find the unknown number.

18. $64 \div 8 = e$

$e =$ ____

19. $0 \div 9 = g$

$g =$ ____

20. ■ $= 20 \div 4$

■ $=$ ____

21. $s = 9 \div 9$

$s =$ ____

MATHEMATICAL PRACTICE 2 **Use Reasoning** **Algebra Complete the table.**

22.

÷	24	40	32	48
8				

23.

÷	54	45	72	63
9				

24. Baseball games have 9 innings. The Little Tigers played 72 innings last season. How many games did the Little Tigers play last year?

____

25. GO DEEPER Sophie has two new fish. She feeds one fish 4 pellets and the other fish 5 pellets each day. If Sophie has fed her fish 72 pellets, for how many days has she had her fish? Explain.

____

____

26. MATHEMATICAL PRACTICE 4 **Write an Equation** Each van going to the aquarium carries 9 students. If 63 third-grade students go to the aquarium, what multiplication fact can you use to find the number of vans that will be needed?

____

Unlock the Problem Real World

27. THINK SMARTER Carlos has 28 blue tang fish and 17 yellow tang fish in one large fish tank. He wants to separate the fish so that there are the same number of fish in each of 9 smaller tanks. How many tang fish will Carlos put in each smaller tank?

a. What do you need to find? ______________________________

____

b. Why do you need to use two operations to solve the problem? ______________________________

____

c. Write the steps to find how many tang fish Carlos will put in each smaller tank.

d. Complete the sentences.

Carlos has ____ blue tang fish and ____ yellow tang fish in one large fish tank.

He wants to separate the fish so that there are the same number of fish in each of ____ smaller tanks.

So, Carlos will put ____ fish in each smaller tank.

28. THINK SMARTER Complete the chart to show the quotients.

÷	27	18	45	36
9				

Name ______________________

Divide by 9

Common Core **COMMON CORE STANDARD—3.OA.C.7**
Multiply and divide within 100.

Find the quotient.

1. $\underline{\ 4\ } = 36 \div 9$
2. $30 \div 6 =$ _____
3. _____ $= 81 \div 9$
4. $27 \div 9 =$ _____
5. $9 \div 9 =$ _____
6. _____ $= 63 \div 7$
7. $36 \div 6 =$ _____
8. _____ $= 90 \div 9$
9. $9\overline{)63}$
10. $9\overline{)18}$
11. $7\overline{)49}$
12. $9\overline{)45}$

Find the unknown number.

13. $48 \div 8 = g$

    $g =$ _____
14. $s = 72 \div 9$

    $s =$ _____
15. $m = 0 \div 9$

    $m =$ _____
16. $54 \div 9 = n$

    $n =$ _____

Problem Solving Real World

17. A crate of oranges has trays inside that hold 9 oranges each. There are 72 oranges in the crate. If all trays are filled, how many trays are there?

    ____

18. Van has 45 new baseball cards. He puts them in a binder that holds 9 cards on each page. How many pages does he fill?

    ____

19. WRITE Math Explain which division facts were the easiest for you to learn.

____

____

Lesson Check (3.OA.C.7)

1. Darci sets up a room for a banquet. She has 54 chairs. She places 9 chairs at each table. How many tables have 9 chairs?

 ____

2. Mr. Robinson sets 36 glasses on a table. He puts the same number of glasses in each of 9 rows. How many glasses does he put in each row?

 ____

Spiral Review (3.OA.A.2, 3.OA.C.7, 3.OA.D.8)

3. Each month for 9 months, Jordan buys 2 sports books. How many more sports books does he need to buy before he has bought 25 sports books?

 ____

4. Find the product.

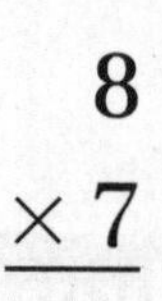

$$\begin{array}{r} 8 \\ \times 7 \\ \hline \end{array}$$

5. Adriana made 30 pet collars to bring to the pet fair. She wants to display 3 pet collars on each hook. How many hooks will Adriana need to display all 30 pet collars?

 ____

6. Carla packs 4 boxes of books. Each box has 9 books. How many books does Carla pack?

 ____

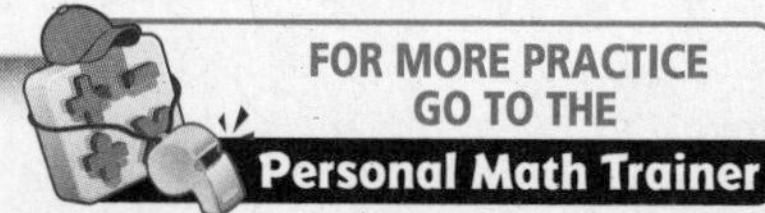

Name ______________________________

Problem Solving • Two-Step Problems

Essential Question How can you use the strategy *act it out* to solve two-step problems?

Common Core **Operations and Algebraic Thinking—3.OA.D.8** *Also 3.OA.A.2, 3.OA.A.3, 3.OA.C.7*

MATHEMATICAL PRACTICES
MP4, MP5, MP6

Unlock the Problem (Real World)

Madilyn bought 2 packs of pens and a notebook for \$11. The notebook cost \$3. Each pack of pens cost the same amount. What is the price of 1 pack of pens?

Read the Problem

What do I need to find?

I need to find the price of

1 pack of __________.

What information do I need to use?

Madilyn spent _____ in all.

She bought _____ packs of pens and _____ notebook.

The notebook cost _____.

How will I use the information?

I will use the information to _____ out the problem.

Solve the Problem

Describe how to act out the problem.

Start with 11 counters. Take away 3 counters.

total cost		cost of notebook		p, cost of 2 packs of pens
↓		↓		↓
____	−	____	=	p
		____	=	p

Now I know that 2 packs of pens cost _____.

Next, make _____ equal groups with the 8 remaining counters.

p, cost of 2 packs of pens		number of packs		c, cost of 1 pack of pens
↓		↓		↓
\$8	÷	____	=	c
		____	=	c

So, the price of 1 pack of pens is _____.

Math Talk MATHEMATICAL PRACTICES 1

Make Sense of Problems Why do you need to use two operations to solve the problem?

Try Another Problem

Chad bought 4 packs of T-shirts. He gave 5 T-shirts to his brother. Now Chad has 19 shirts. How many T-shirts were in each pack?

Read the Problem	Solve the Problem
What do I need to find?	**Describe how to act out the problem.**
What information do I need to use?	
How will I use the information?	

- How can you use multiplication and subtraction to check your answer?

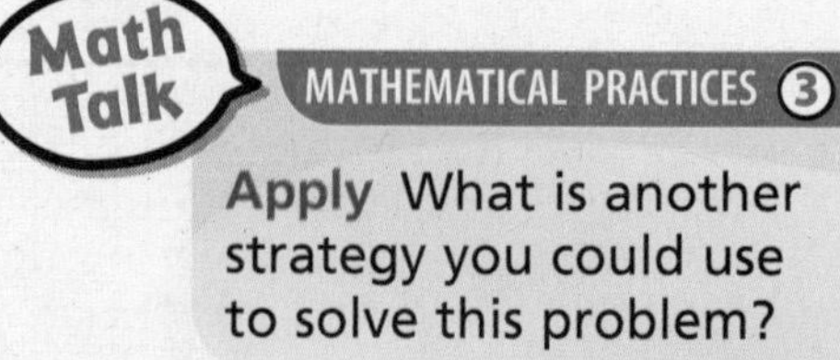

Name ___________________________

Share and Show

Unlock the Problem

- √ Circle the question.
- √ Underline the important facts.
- √ Choose a strategy you know.

1. Mac bought 4 packs of toy cars. Then his friend gave him 9 cars. Now Mac has 21 cars. How many cars were in each pack?

Act out the problem by using counters or the picture and by writing equations.

First, subtract the cars Mac's friend gave him.

total cars ↓		cars given to Mac ↓		c, cars in 4 packs ↓
21	−	____	=	c
		____	=	c

Then, divide to find the number of cars in each pack.

c, cars in 4 packs ↓		number of packs ↓		p, number in each pack ↓
12	÷	____	=	p
		____	=	p

So, there were ____ cars in each pack.

2. THINKSMARTER What if Mac bought 8 packs of cars and then he gave his friend 3 cars? If Mac has 13 cars now, how many cars were in each pack?

____

On Your Own

3. THINKSMARTER Ryan gave 7 of his model cars to a friend. Then he bought 6 more cars. Now Ryan has 13 cars. How many cars did Ryan start with?

____

4. GO DEEPER Chloe bought 5 sets of books. She donated 9 of her books to her school. Now she has 26 books. How many books were in each set?

____

5. Hilda cuts a ribbon into 2 equal pieces. Then she cuts 4 inches off one piece. That piece is now 5 inches long. What was the length of the original ribbon?

____

6. GO DEEPER Teanna has 2 boxes of color pencils. One box has 20 color pencils and the other box has 16 color pencils. She gives her brother 3 of the color pencils. She wants to put the color pencils that she has left into 3 equal groups. How many color pencils can Teanna put in each group?

____

WRITE Math
Show Your Work

7. MATHEMATICAL PRACTICE 6 Rose saw a movie, shopped, and ate at a restaurant. She did not see the movie first. She shopped right after she ate. In what order did Rose do these activities? **Explain** how you know.

____

____

____

____

____

Personal Math Trainer

8. THINKSMARTER+ Eleni bought 3 packs of crayons. She then found 3 crayons in her desk. Eleni now has 24 crayons. How many crayons were in each pack she bought? Explain how you solved the problem.

____

____

Name ______________________

Problem Solving • Two-Step Problems

COMMON CORE STANDARD—3.OA.D.8
Solve problems involving the four operations, and identify and explain patterns in arithmetic.

Solve the problem.

1. Jack has 3 boxes of pencils with the same number of pencils in each box. His mother gives him 4 more pencils. Now Jack has 28 pencils. How many pencils are in each box?
   Think: I can start with 28 counters and act out the problem.

   8 pencils

2. The art teacher has 48 paintbrushes. She puts 8 paintbrushes on each table in her classroom. How many tables are in her classroom?

   ____

3. Ricardo has 2 cases of video games with the same number of games in each case. He gives 4 games to his brother. Ricardo has 10 games left. How many video games were in each case?

   ____

4. Patty has $20 to spend on gifts for her friends. Her mother gives her $5 more. If each gift costs $5, how many gifts can she buy?

   ____

5. Joe has a collection of 35 DVD movies. He received 8 of them as gifts. Joe bought the rest of his movies over 3 years. If he bought the same number of movies each year, how many movies did Joe buy last year?

   ____

6. **WRITE** *Math* Write a division word problem and explain how to solve it by *acting it out*.

   ____

   ____

Lesson Check (3.OA.D.8)

1. Gavin saved $16 to buy packs of baseball cards. His father gives him $4 more. If each pack of cards costs $5, how many packs can Gavin buy?

2. Chelsea buys 8 packs of markers. Each pack contains the same number of markers. Chelsea gives 10 markers to her brother. Then, she has 54 markers left. How many markers were in each pack?

Spiral Review (3.OA.A.1, 3.OA.A.3, 3.OA.A.4, 3.OA.D.8)

3. Each foot has 5 toes. How many toes do 6 feet have?

4. Each month for 5 months, Sophie makes 2 quilts. How many more quilts does she need to make before she has made 16 quilts?

5. Meredith practices the piano for 3 hours each week. How many hours will she practice in 8 weeks?

6. Find the unknown factor.

$9 \times ■ = 36$

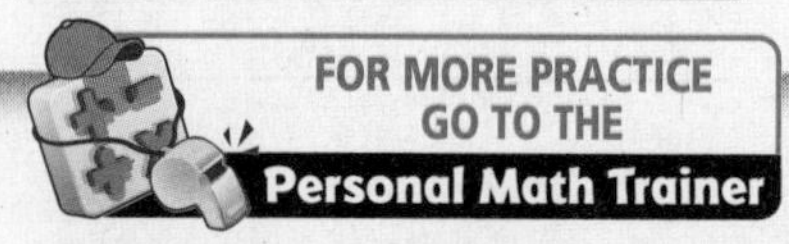

Name ____________________

Order of Operations

Essential Question Why are there rules such as the order of operations?

Common Core **Operations and Algebraic Thinking—3.OA.D.8** *Also 3.OA.A.1, 3.OA.A.2, 3.OA.A.3, 3.OA.C.7*

MATHEMATICAL PRACTICES
MP2, MP4, MP6

Investigate

CONNECT You can use what you know about acting out a two-step problem to write one equation to describe and solve a two-step problem.

- If you solved a two-step problem in a different order, what do you think might happen?

____

Use different orders to find $4 + 16 \div 2$.

A. Make a list of all the possible orders you can use to find the answer to $4 + 16 \div 2$.

____

B. Use each order in your list to find the answer. Show the steps you used.

____

____

Draw Conclusions

1. Did following different orders change the answer? ____________________

2. MATHEMATICAL PRACTICE 8 **Draw Conclusions** If a problem has more than one type of operation, how does the order in which you perform the operations affect the answer?

____

3. Explain the need for setting an order of operations that everyone follows.

____

____

Make Connections

When solving problems with more than one type of operation, you need to know which operation to do first. A special set of rules, called the **order of operations**, gives the order in which calculations are done in a problem.

First, multiply and divide from left to right.

Then, add and subtract from left to right.

Meghan buys 2 books for \$4 each. She pays with a \$10 bill. How much money does she have left?

You can write $\$10 - 2 \times \$4 = c$ to describe and solve the problem.

Use the order of operations to solve $\$10 - 2 \times \$4 = c$.

STEP 1

Multiply from left to right.

$\$10 - 2 \times \$4 = c$

$\$10 - \quad \$8 \quad = c$

STEP 2

Subtract from left to right.

$\$10 - \$8 = c$

$\$2 \quad = c$

So, Meghan has _____ left.

- Does your answer make sense? Explain.

____

____

____

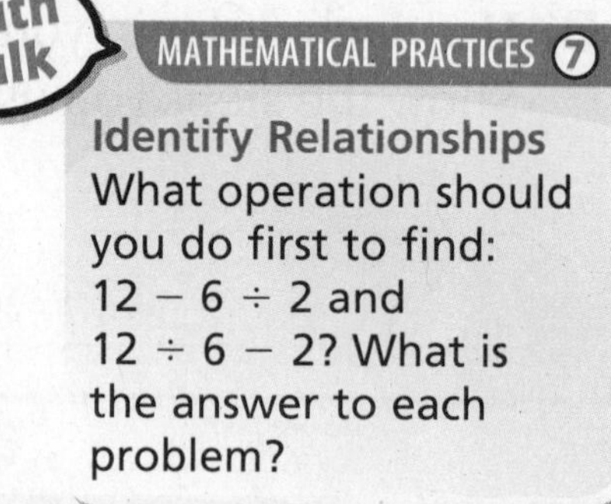

Identify Relationships
What operation should you do first to find: $12 - 6 \div 2$ and $12 \div 6 - 2$? What is the answer to each problem?

Write *correct* if the operations are listed in the correct order. If not correct, write the correct order of operations.

1. $4 + 5 \times 2$ multiply, add

____

2. $8 \div 4 \times 2$ multiply, divide

____

3. $12 + 16 \div 4$ add, divide

____

4. $9 + 2 \times 3$ add, multiply

____

5. $4 + 6 \div 3$ divide, add

____

6. $36 - 7 \times 3$ multiply, subtract

____

Name ______________________

Follow the order of operations to find the unknown number. Use your MathBoard.

7. $63 \div 9 - 2 = f$

$f =$ _____

8. $7 - 5 + 8 = y$

$y =$ _____

9. $3 \times 6 - 2 = h$

$h =$ _____

10. $80 - 64 \div 8 = n$

$n =$ _____

11. $3 \times 4 + 6 = a$

$a =$ _____

12. $2 \times 7 \div 7 = c$

$c =$ _____

Problem Solving • Applications

MATHEMATICAL PRACTICE 4 Write an Equation **Algebra** **Use the numbers listed to make the equation true.**

13. 2, 6, and 5

_____ + _____ × _____ = 16

14. 4, 12, and 18

_____ − _____ ÷ _____ = 15

15. 8, 9, and 7

_____ × _____ − _____ = 47

16. 2, 4, and 9

_____ ÷ _____ + _____ = 11

17. WRITE Math **Pose a Problem** Write a word problem that can be solved by using $2 \times 5 \div 5$. Solve your problem.

____

____

____

____

18. THINK SMARTER Is $4 + 8 \times 3$ equal to $4 + 3 \times 8$? Explain how you know without finding the answers.

____

____

____

19. THINK SMARTER For numbers 19a–19d, select True or False for each equation.

19a. $24 \div 3 + 5 = 13$ ○ True ○ False

19b. $5 + 2 \times 3 = 21$ ○ True ○ False

19c. $15 - 3 \div 3 = 14$ ○ True ○ False

19d. $18 \div 3 \times 2 = 12$ ○ True ○ False

Connect to Social Studies

Picture Book Art

The Eric Carle Museum of Picture Book Art in Amherst, Massachusetts, is the first museum in the United States that is devoted to picture book art. Picture books introduce literature to young readers.

The museum has 3 galleries, a reading library, a café, an art studio, an auditorium, and a museum shop. The exhibits change every 3 to 6 months, depending on the length of time the picture art is on loan and how fragile it is.

Souvenir Prices	
Souvenir	**Price**
Firefly Picture Frame	$25
Exhibition Posters	$10
Caterpillar Note Cards	$8
Caterpillar Pens	$4
Sun Note Pads	$3

The table shows prices for some souvenirs in the bookstore in the museum.

20. Kallon bought 3 Caterpillar note cards and 1 Caterpillar pen. How much did he spend on souvenirs?

____

21. GO DEEPER Raya and 4 friends bought their teacher 1 Firefly picture frame. They shared the cost equally. Then Raya bought an Exhibition poster. How much money did Raya spend in all? Explain.

____

____

____

Name ______________________________

Order of Operations

COMMON CORE STANDARD—3.OA.D.8
Solve problems involving the four operations, and identify and explain patterns in arithmetic.

Write *correct* if the operations are listed in the correct order. If not correct, write the correct order of operations.

1. $45 - 3 \times 5$ subtract, multiply

   multiply, subtract

2. $3 \times 4 \div 2$ divide, multiply

   ____

3. $5 + 12 \div 2$ divide, add

   ____

4. $7 \times 10 + 3$ add, multiply

   ____

Follow the order of operations to find the unknown number.

5. $6 + 4 \times 3 = n$

   $n =$ ______

6. $8 - 3 + 2 = k$

   $k =$ ______

7. $24 \div 3 + 5 = p$

   $p =$ ______

Problem Solving Real World

8. Shelley bought 3 kites for $6 each. She gave the clerk $20. How much change should Shelley get?

   ____

9. Tim has 5 apples and 3 bags with 8 apples in each bag. How many apples does Tim have in all?

   ____

10. WRITE Math Give a description of the rules for the order of operations in your own words.

   ____

   ____

Lesson Check (3.OA.D.8)

1. Natalie is making doll costumes. Each costume has 4 buttons that cost 3¢ each and a zipper that costs 7¢. How much does she spend on buttons and a zipper for each costume?

2. Leonardo's mother gave him 5 bags with 6 flower bulbs in each bag to plant. He has planted all except 3 bulbs. How many flower bulbs has Leonardo planted?

Spiral Review (3.OA.C.7, 3.OA.D.9, 3.NBT.A.3)

3. Each story in Will's apartment building is 9 feet tall. There are 10 stories in the building. How tall is the apartment building?

4. Describe pattern in the table.

Tables	1	2	3	4
Chairs	4	8	12	16

5. For decorations, Meg cut out 8 groups of 7 snowflakes each. How many snowflakes did Meg cut out in all?

6. A small van can hold 6 students. How many small vans are needed to take 36 students on a field trip to the music museum?

FOR MORE PRACTICE GO TO THE **Personal Math Trainer**

Name ________________________________

Chapter 7 Review/Test

1. Ming divided 35 marbles between 7 different friends. Each friend received the same number of marbles. How many marbles did Ming give to each friend?

$$35 \div 7 = a$$
$$7 \times a = 35$$

Ⓐ 4

Ⓑ 5

Ⓒ 6

Ⓓ 7

2. Mrs. Conner has 16 shoes.

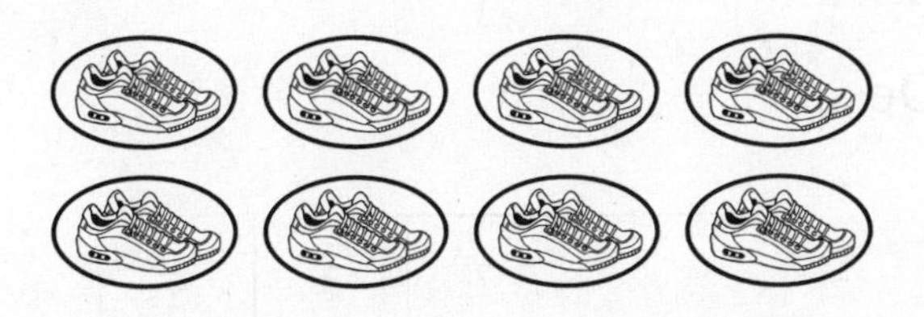

Select one number from each column to show the division equation represented by the picture.

$16 \div \underset{\text{(divisor)}}{\underline{?}} = \underset{\text{(quotient)}}{\underline{?}}$

Divisor	Quotient
○ 1	○ 1
○ 2	○ 4
○ 4	○ 8
○ 16	○ 16

3. Twenty boys are going camping. They brought 5 tents. An equal number of boys sleep in each tent. How many boys will sleep in each tent?

__________ boys

GO DIGITAL Assessment Options Chapter Test

4. Circle a number for the unknown factor and quotient that makes the equation true.

	6			6	
$4 \times$	7	$= 28$		7	$= 28 \div 4$
	8			8	

5. Mrs. Walters has 30 markers. She gives each student 10 markers. How many students received the markers?

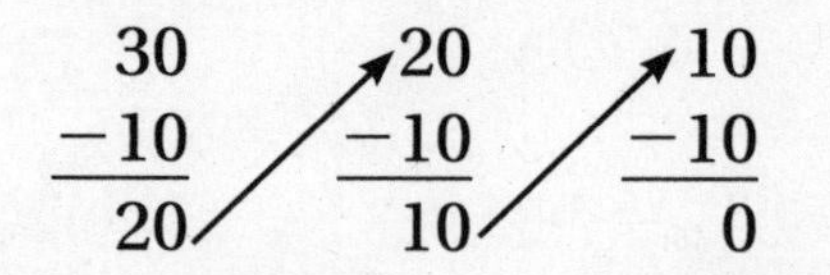

Write a division equation to represent the repeated subtraction.

______ ÷ ______ = ______

6. Complete the chart to show the quotients.

÷	27	36	45	54
9				

7. For numbers 7a–7e, select True or False for each equation.

7a. $12 \div 6 = 2$ ○ True ○ False

7b. $24 \div 6 = 3$ ○ True ○ False

7c. $30 \div 6 = 6$ ○ True ○ False

7d. $42 \div 6 = 7$ ○ True ○ False

7e. $48 \div 6 = 8$ ○ True ○ False

Name ______________________________

8. Alicia says that $6 \div 2 + 5$ is the same as $5 + 6 \div 2$. Is Alicia correct or incorrect? Explain.

____

____

____

____

____

9. Keith arranged 40 toy cars in 8 equal rows. How many toy cars are in each row?

__________ toy cars

10. Bella made $21 selling bracelets. She wants to know how many bracelets she sold. Bella used this number line.

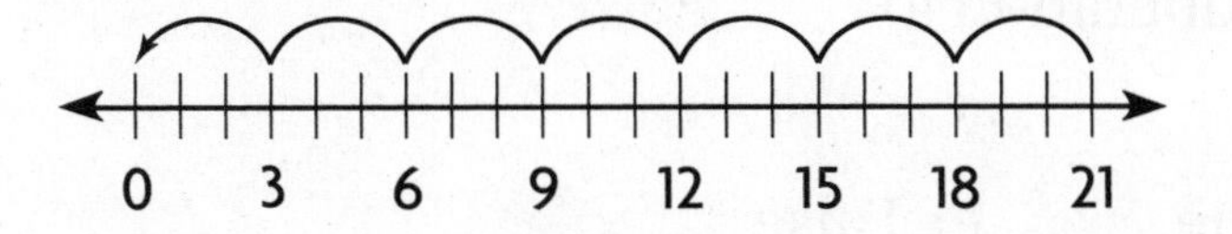

Write the division equation that the number line represents.

_________ $\div$ _________ $=$ _________

11. Each picnic table seats 6 people. How many picnic tables are needed to seat 24 people? Explain the strategy you used to solve the problem.

____

____

____

12. Finn bought 2 packs of stickers. Each pack had the same number of stickers. A friend gave him 4 more stickers. Now he has 24 stickers in all. How many stickers were in each pack? Explain how you solved the problem.

____

____

____

13. Ana used 49 strawberries to make 7 strawberry smoothies. She used the same number of strawberries in each smoothies. How many strawberries did Ana use in each smoothie?

________ strawberries

14. For numbers 14a–14e, use the order of operation to select True or False for each equation.

14a.	$81 \div 9 + 2 = 11$	○ True	○ False
14b.	$6 + 4 \times 5 = 50$	○ True	○ False
14c.	$10 + 10 \div 2 = 15$	○ True	○ False
14d.	$12 - 3 \times 2 = 6$	○ True	○ False
14e.	$20 \div 4 \times 5 = 1$	○ True	○ False

15. A flower shop sells daffodils in bunches of 9. It sells 27 daffodils. How many bunches of daffodils does the shop sell?

________ bunches

Name ___________________________

Personal Math Trainer

16. THINKSMARTER+ Aviva started a table showing a division pattern.

÷	20	30	40	50
10				
5				

Part A

Complete the table.

Compare the quotients when dividing by 10 and when dividing by 5. Describe a pattern you see in the quotients.

____

____

Part B

Find the quotient, a.

$70 \div 10 = a$

$a =$ ___________

How could you use a to find the value of n? Find the value of n.

$70 \div 5 = n$

$n =$ ___________

____

____

17. Ben needs 2 oranges to make a glass of orange juice. If oranges come in bags of 10, how many glasses of orange juice can he make using one bag of oranges?

___________ glasses

18. For numbers 18a–18e, select True or False for each equation.

18a. $0 \div 9 = 0$ ○ True ○ False

18b. $9 \div 9 = 1$ ○ True ○ False

18c. $27 \div 9 = 4$ ○ True ○ False

18d. $54 \div 9 = 6$ ○ True ○ False

18e. $90 \div 9 = 9$ ○ True ○ False

19. Ellen is making gift baskets for four friends. She has 16 prizes she wants to divide equally among the baskets. How many prizes should she put in each basket?

______ prizes

20. GO DEEPER Emily is buying a pet rabbit. She needs to buy items for her rabbit at the pet store.

Part A

Emily buys a cage and 2 bowls for $54. The cage costs $40. Each bowl costs the same amount. What is the price of 1 bowl? Explain the steps you used to solve the problem.

______

______

______

Part B

Emily also buys food and toys for her rabbit. She buys a bag of food for $20. She buys 2 toys for $3 each. Write one equation to describe the total amount Emily spends on food and toys. Explain how to use the order of operations to solve the equation.

______

______

______